MARKETING FOR FARM

AND

RURAL ENTERPRISE

MARKETING FOR FARM

AND

RURAL ENTERPRISE

MICHAEL HAINES

FARMING PRESS

FARMING PRESS

First published 1999

Copyright © Michael Haines 1998
1 3 5 7 9 10 8 6 4 2

All rights reserved. No part of this
publication may be reproduced, stored
in a retrieval system, or transmitted,
in any form or by any means, electronic,
mechanical, photocopying, recording or otherwise,
without prior permission of Farming Press.

ISBN 0 85236 405 9

A catalogue record for this book is
available from the British Library

**Published by Farming Press
Miller Freeman UK Ltd
2 Wharfedale Road, Ipswich, IP1 4LG
United Kingdom**

Distributed in North America by
Diamond Farm Enterprises,
Box 537, Bailey Settlement Road,
Alexandria Bay, NY13607, USA

Cover design by Ian Garstka
Printed and bound in Great Britain by Biddles, Guildford, Surrey

CONTENTS

Chapter 1: Introduction: why marketing matters	1
Chapter 2: What marketing means	7
Chapter 3: Analysing the current business	15
Chapter 4: Scanning the marketing environment	33
Chapter 5: Understanding the marketing system	51
Chapter 6: Analysing the market	71
Chapter 7: The product decision	95
Chapter 8: The price decision	115
Chapter 9: The place decision	135
Chapter 10: Selling the product	153
Chapter 11: Promotion	169
Chapter 12: Group marketing	187
Chapter 13: Export marketing	201
Chapter 14: Implementation and control	217
Glossary	223
Index	228

PICTURE CREDITS

Plates 1, 3, 6, 11: *The Grocer*
Plates 2, 15: Swedish Farmers UK Ltd
Plate 4: J Sainsbury Plc
Plates 5, 8: Safeway
Plates 7a, 7b: Rachel's Dairy Ltd
Plates 9, 10, 18: Meat & Livestock Commission
Plates 12, 13, 14, 16, 19, 21: Welsh Food Promotions Ltd
Plates 17a, 17b: *Farmers Weekly*
Plate 20: Development Board for Rural Wales

FOREWORD

This book provides an introduction to marketing in the context of the farm-based business as it finds itself increasingly exposed to market forces.

It is management-oriented, not theoretical, introducing marketing concepts and terminology only as they are relevant to the owner-managed business. It is intended both as an introductory student text, and one which will also be useful to farmers who need to improve their competitive position and their revenue from the market. For managers of diversified farm businesses, it will formalise what they have learned about marketing in practice, and provide a basis for more strategic planning and improved performance.

The existence of many other marketing textbooks may prompt the question why another one is needed for the farm business. The answer is that although the principles of marketing do not vary from one business to another, their application is severely constrained in the farm-based business because it is so subject to political intervention. Marketing textbooks are therefore a frustrating read for the farm business manager, because his options are much less simple than the books suggest. This one considers only what is feasible in a farming and rural enterprise, drawing on many years of firsthand management and advisory experience in real farm business situations.

I am grateful to Paul King and Alan Leather for the preparation of graphic material, and to Hal Norman of the Farming Press for seeing the book through to publication.

Michael Haines
Aberystwyth, December 1998

CHAPTER 1

Introduction: why marketing matters

Marketing is about improving revenue and achieving a competitive advantage in the market, by better targeting of customers and more effective marketing strategies.

For as long as most farmers have been in business this has largely been irrelevant, because it has been possible to make a reasonable living from farming with scarcely a thought for customers or competitors. In most countries the farmer's only customer for most agricultural commodities has effectively been the government, which provided a product specification and underwrote production with support prices only marginally linked (if that) to real customer requirements. The marketing decisions which other business managers take for granted were transferred to politicians and bureaucrats, while farmers were encouraged to become production specialists – producing first-class products to specifications which ignored demand in real markets, at a cost governments can no longer underwrite.

In the process farmers have been isolated from the consumer and from the downstream sector which transforms their output into consumer products. They have also been shielded from the full impact of commercial competition, including competition with other farmers. It is therefore not surprising that most farmers are little more than raw material suppliers, who have only minimal contact with the customer, minimal control over the movement of their output through the marketing system, and minimal opportunity for adding value. Too often their only direct marketing responsibility is the production decision and quality control. These are both vital marketing functions, but neither takes marketing very far.

This situation is increasingly untenable in the face of:

- declining agricultural price support
- output controls in many sectors
- intensifying competition for markets
- consumer demand for product assurance.

It is only tenable now because agricultural price support has been replaced by other direct and indirect income supports: for example, disadvantaged area and countryside management payments. However, in most countries the total cost of subsidies and other payments to farmers is now so large that budgetary pressure will inevitably force further cutbacks. For farmers who wish to reduce their dependence on the political process and increase their return from the market, the need for better marketing has therefore never been more genuine nor more urgent.

The support net disappeared years ago in some countries (notably New Zealand), exposing farmers to the full blast of market forces and the need to become competitive in world markets. In Europe the process started with the 1992 MacSharry reform of the CAP, but the impact of the reform on farmers was reduced because farm incomes were artificially sustained through other support mechanisms, by currency shifts and a short-lived boost in world prices. Reform of the CAP is nevertheless inevitable, as more countries with large agricultural sectors join the EU and the taxpayer burden escalates.

The 1994 GATT world trade agreement also committed signatories to the progressive elimination of direct subsidisation of agricultural exports, which is another form of producer subsidy. The implementation of the agreement over the period 1995–99 is moving agriculture progressively towards the marketplace, and producer prices are increasingly determined by the world market, not by politicians (although politicians will continue to restrict output and intervene with direct and indirect income supports). Market returns will therefore increasingly reflect world prices, and anything more than a floor price will depend on better marketing, since production restraints in most sectors have closed the traditional option of expanding output to maintain profitability.

THE BUSINESS ENVIRONMENT

Alongside this process of reform, global competition is intensifying for virtually static food markets in most developed countries. It may still be true that there is a market somewhere at some price for most products, but quality-conscious consumers in developed economies expect a good product and a competitive price together with added services and value. Consumer mistrust of agricultural and food production methods has also been reinforced by media scares which have deeply undermined confidence in the product. The result is that consumers increasingly dictate the terms of supply not just for the finished product, but also for the way in which it is produced. Producers who cannot or will not meet this demand will therefore find it increasingly difficult to find a market at any price.

New handouts to 'complying' producers

Brussels is working on a blueprint for reform of the CAP which will abolish export subsidies for major foods, intervention and production-linked subsidies for farmers. In its place a 'new rural policy' is planned, which will take subsidies away from food production and use them to pay farmers to stop farming, to preserve villages and to prevent the drift from the countryside ... Present prices would be rapidly scaled down to the point where they would be equal to world price levels. This would eliminate the need for either export subsidies on dairy products, meat, cereals or sugar products ... for some products this would mean a cut in returns by as much as 30%. Farmers would be cushioned by the payment of direct subsidies equal to the income drop. The subsidies would however be completely detached from the farmer's output and ... producers would not get subsidies unless they 'complied' with new EU rules on environment-friendly farming practices.

The Grocer, 30 November 1996

Box 1.1

WHY MARKETING MATTERS

The expansion of world population will admittedly provide many more mouths to feed in the Third World, but it is not clear how they will pay for the food. Marketing opportunities may also arise if incomes and expectations continue to rise in Asia and the expanding economies of Eastern Europe (CEEC) and the former USSR. However, these countries have large developing farm sectors of their own which will progressively supply their own needs and present new competition in world markets. Meanwhile, highly efficient and cost-effective suppliers in Australia and New Zealand, South America and the USA are queuing up to fill the gap.

The immediate reality for most farmers is in other words a buyer's market in which food manufacturers, retailers and consumers can take their pick from worldwide suppliers, with ruthless disregard for any who fail to make the grade. Preserving existing markets, let alone winning new ones, therefore depends on gaining a competitive advantage by better marketing. Better marketing also holds the key to improved farm profitability, by providing opportunities to win a larger share of consumer food expenditure through added value and greater customer satisfaction. Production efficiency and further cost reduction will continue to be important in maintaining farm incomes, but this may only be achievable by increases in farm size which bring economies of scale. Where this is not an option, and there is not much fat left in the production system, the only area with any significant room for improvement is marketing.

THE MARKETING OPPORTUNITY

The opportunity for better marketing lies in the fact that barely a quarter of farm output reaches the consumer in the state in which it left the farm. Most farmers know this, but they underestimate the marketing effort and the value added by the downstream marketing sector, and the missed business opportunity which this represents.

Most farm output is in fact useless to the consumer without some kind of processing and intermediary service. It is highly perishable, and sometimes even alive; it is not standardised by size or quality; it is variable in quantity and quality over the year, and produced far from centres of consumption. Processing and distribution services add utility and value to the raw material by transforming it into consistently available consumer products, and in the process provide a living for the firms who make this their business.

In addition to covering their costs these firms clearly need to earn a margin on their operations and capital investment, and it is their profit and

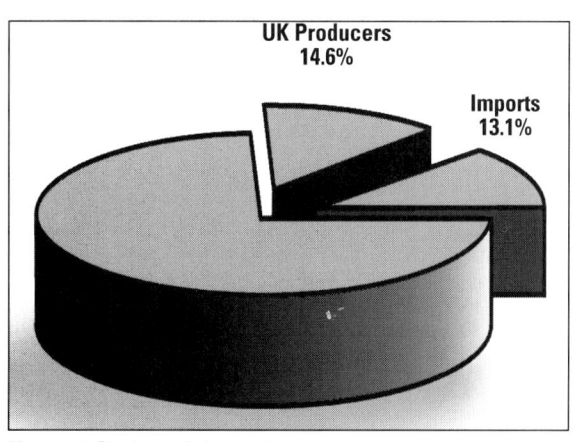

Figure 1.1 Producers' share of consumer expenditure on food and drink in 1995

UK Producers 14.6%

Imports 13.1%

costs which largely explain the gap between producer and consumer prices. In other words, if the farmer's share of consumer food expenditure is currently small (Figure 1.1) it is because his contribution to the satisfaction of consumer needs is currently limited, and a larger share will depend on:

- providing services currently supplied by the downstream sector: for example, by processing farm output or selling direct to the public
- achieving a marketing premium by contributing to improved channel efficiency: for example, by reducing buyers' procurement and processing costs, or meeting a specification they cannot otherwise reliably obtain.

On-farm processing and direct sale

The possibility of processing farm output and/or selling direct to the public appeals to many producers as a way of increasing revenue, and on a small scale it may be achievable because existing facilities and spare labour can be utilised. However, the return from the additional management burden and investment rarely achieves farmers' expectations because adding value always adds costs, but does not always bring higher revenue. Scaling up the enterprise also entails extra capital and labour, together with new technical and marketing expertise, and this may exceed the resources of the small owner-managed business. The resulting financial and management commitment can become a burden, and if it diverts attention and capital from other parts of the business it may adversely affect total business performance and revenue. Before embarking on this route there must therefore be a reasonable certainty that sufficient revenue will be generated to cover the added costs and yield a worthwhile profit; that the development is sustainable over a reasonably long term, and that total business revenue will not suffer.

For most producers, better marketing through conventional marketing channels is a more realistic option. It also achieves the main benefit to be derived from added-value enterprises – which is not profit, but better customer feedback and greater influence over the post-farmgate use and destination of output. However, this requires a better understanding of the marketing system than most farmers possess, and a much greater willingness to cooperate with the downstream sector and with other producers.

Better marketing through existing channels

Experience shows that both manufacturers and retailers will pay a premium for genuine added value in the form of product quality and benefits like out-of-season production, part-processing, or special delivery schedules. In sectors such as milk and grain, producers are accustomed to work to such payment incentives. The opportunities extend far beyond these sectors, however, and changes in the marketing environment will inevitably expand them. Quota restriction of some raw materials (most obviously milk) has already put a premium on secure channels of supply for manufacturing and retailing. Food technology and statutory hygiene requirements, together with customer demand for traceability (guaranteed quality and production standards) have also reinforced the need for inputs of guaranteed, consistent quality.

Retailers in particular are anxious to establish marketing relationships and formal producer alliances which shorten the marketing channel and ensure traceability (for examples, see Plates 1 and 9). Many retailers are also keen to supply locally-sourced and farm-processed products, provided these can be integrated into their sophisticated distribution systems without compromising their own cost objectives or their ability to comply with rigorous food hygiene requirements.

Many producers are rightly wary of locking themselves into such marketing relationships without a better understanding of the benefits and risks, and the management implications. It is also true that the effort may not always attract a price premium, but price penalties are increasingly imposed for uncompetitive products. Recurrent media scares in the 1990s have also finally convinced many farmers that secure access to the market may be as important as price.

THE MARKETING CHALLENGE

The success with which some farmers have diversified into alternative enterprises has proved beyond doubt that they are perfectly capable of marketing their output effectively in a free market situation, once they have the incentive. Agricultural markets are still not free, however, and support policies still act as a disincentive to better marketing.

There is also less scope for individual, entrepreneurial marketing of mainstream agricultural commodities. This is true of all raw materials because they are largely undifferentiated, and they are bought in bulk by industrial buyers whose primary interest is to reduce input costs. Raw output is consequently difficult to brand (establish an identity distinct from competitor output). Branding is an essential condition for customer recognition and repeat sales, however, and the ability to establish a brand is a prerequisite for most marketing activity. Value added activities also tend to require large volumes of processing material, to justify and produce a return on the necessary capital investment. In most developed countries the domination of the market by a few massive retail and manufacturing companies further disadvantages the lone producer. Joint ventures with other producers may therefore be the only way to supply a branded product, to add value to commodities, and to deliver the year-round volume of high-specification which the market demands – which is why this book has a chapter on group marketing.

Even with better marketing, however, it is likely that many farmers will have difficulty in maintaining their income from the market without diversifying into new products and enterprises. Diversification opportunities are not unlimited, but some will always exist for entrepreneurs who have the initiative to spot them and the dedication and the resources to exploit them. Consumer demand continues to expand for more 'natural' and more locally-sourced food products which retailers are anxious to supply and farmers are well placed to provide. Demand also exists for more recreational and environmental amenities which farmers can supply. Some farmers may still also benefit from rural relocation of light industry and commerce, as deteriorating urban conditions and improving communications technology encourage more people to work in the countryside.

In time the diversified farm business may therefore become the rule rather than the exception, but the evidence is that many diversified farm enterprises have not achieved their financial promise, and many do not break even. Better marketing is therefore likely to be the key to profitable diversification, as it is the key to sustainable farming viability.

CONCLUSION

There are no magic formulas for better marketing, but this book provides an introduction which will at least make it more readily identifiable in the form of management strategies and action appropriate to the owner-managed business.

Chapter 2 suggests how better marketing can be built into the farm planning routine without substantially increasing the management burden. Chapters 3–6 outline the information required for sound marketing decisions, and Chapters 7–11 consider the main marketing decision areas: product, price, distribution and promotion. Chapters 12–14 discuss group marketing and export marketing, and the importance of effective implementation and control, without which even the best planning is unlikely to succeed.

CHAPTER 2

What marketing means

The plea for better marketing is frequently ignored because many managers think they are already marketing their output effectively – by which they mean disposing profitably of what they happen to have produced. This confuses marketing with *selling*, though as Figure 2.1 indicates, selling is only part of the marketing process, and in a free market rarely compensates for producing the wrong product, at the wrong price, at the wrong time and place. Many managers also think marketing is synonymous with advertising and 'the hard sell', which they associate with aggressive salesmanship and an inferior product. It is therefore not surprising that they are offended by the suggestion that their output needs better marketing, believing that a 'good product sells itself'.

In markets where demand exceeds supply (because of absolute shortage or an innovative product in short supply) it may still be true that a good product sells itself. One of the objectives of better marketing is indeed to spot such opportunities ahead of the competition, in order to achieve any price premium associated with short supply. However, in oversupplied markets with sophisticated consumers, it is rarely true that a

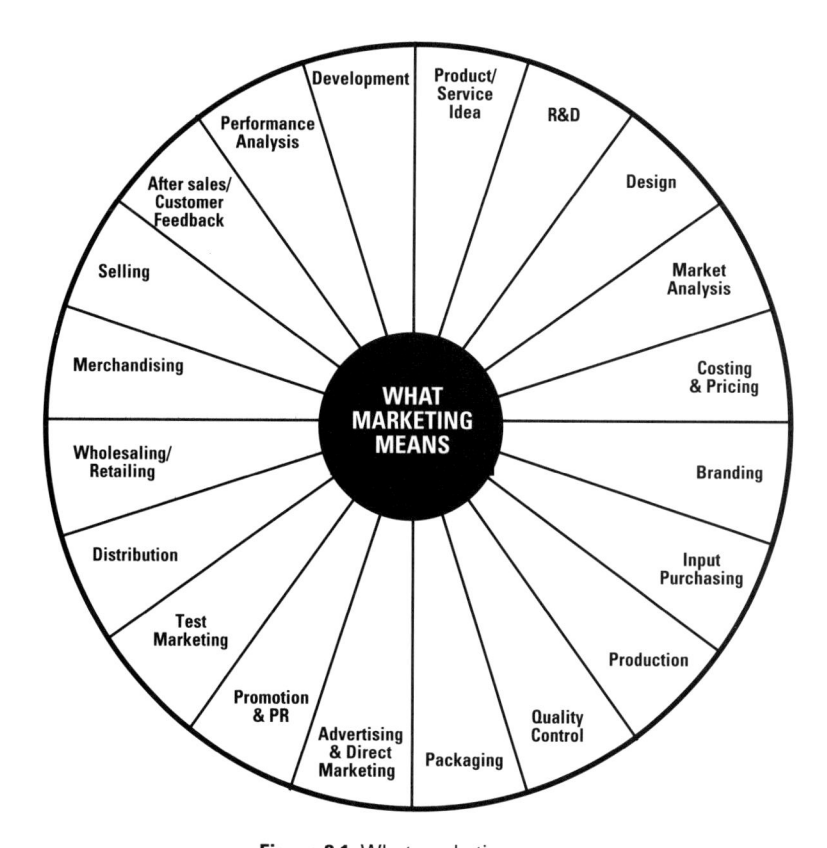

Figure 2.1 What marketing means

good product sells itself, or that clever advertising and sharp salesmanship will achieve repeat sales of a poor product. It is certainly not true of food markets in developed economies, where comparable, competitively priced products must compete for consumer expenditure by adding value in the form of product assurance, convenience, variety, presentation, status, fashionability – even a clear customer conscience.

Marketing is a management approach which sets out actively to maximise the potential for added value, with the intention of gaining a competitive advantage and any price premium available for a superior product. As Figure 2.1 indicates, this entails a complete management package which *includes* selling and advertising, but in conjunction with many other product-related activities. Marketing is not, in other words, a bolt-on option which can be added to the production process: marketing *directs* production from start to finish.

As one memorable definition puts it, marketing starts with the decision to produce a *saleable* product (OECD, 1964). This apparently circuitous statement stresses the fact that in unsubsidised markets, revenue depends on real sales to satisfied customers. Customer satisfaction – not technical efficiency or financial performance – is therefore the primary test of business performance. Since everyone else in the same line of business also knows this, and is seeking to win sales by maximising customer satisfaction, it also follows that marketing is about beating the competition.

Beating the competition means researching and identifying a customer need which a business either does or could supply, and delivering it more effectively than competitor offers. This may entail product modifications: for example, low-fat or organic milk; en suite farmhouse accommodation with swimming pool. Sometimes no production changes are required, but only added services which enhance a standard offer: washed and graded vegetables; single-portion and family packs; farmhouse accommodation with baby sitting service. Sometimes neither production changes nor added services are necessary, but only better distribution and/or efforts to change customer perceptions of the product: farm assured, 'new improved', healthy, humane, environmentally-friendly.

Generally, product modifications, better distribution and efforts to change customer perceptions go hand in hand, since there is no point in improving products if customers do not perceive the change, and improved customer perceptions will not bring repeat sales if a product fails to live up to the promise. There is also no point in adding value unless added costs are recovered and a reasonable return is made on the additional capital investment and management effort. The strategy must also be consistent with overall business objectives.

A complete definition of marketing therefore includes all the following components:

a management approach which seeks actively to maximise customer satisfaction and establish a competitive advantage by supplying a well-defined customer need, at a price acceptable to the customer and sufficient to achieve business objectives and long-run profitability.

In practice this is difficult to achieve because:

- customer needs are not easy to identify precisely and are constantly changing

- the competition is difficult to identify and even more difficult to assess, in the absence of reliable information about commercial competitors
- the information and planning requirement may exceed the management resources of an owner-managed business
- marketing costs money, and these costs must be built into acceptable budgets related to realistic sales targets
- all these decision areas interact, causing new management problems and complicating the decision-making process
- external factors constrain management decisions and control.

MARKETING MANAGEMENT

Marketing management is the term normally applied to the systematic analysis of these problems, the identification and evaluation of alternative marketing options, and the design and implementation of an effective marketing programme.

For farmers used to the discipline of farm management planning this will not sound unfamiliar. Indeed, the problem is that it may seem too familiar to merit all the fuss. Unlike many new skills which farmers have had to master, marketing is not a radically new departure requiring radically new techniques. It is simply a management re-orientation which ensures that business activity is *demand-led*, not *production-driven*: a phrase often heard but rarely understood. It does not downgrade the importance of technical and production efficiency or financial performance, which obviously remain fundamental to business viability. It simply shifts the management focus by insisting that planning and production be guided primarily by *marketing* objectives and requirements (Table 2.1).

IN THE PRODUCTION-ORIENTED BUSINESS:

- 'Marketing' means disposing of what happens to have been produced
- The management focus and emphasis are on production
- Products are 'over-engineered' to satisfy own standards, regardless of customer requirements or willingness to pay
- Marketing research and planning are almost non-existent
- Price tends to be cost-based, with value and competitive considerations largely ignored
- Cost reduction efforts dominate, and may sacrifice product quality and customer service
- Instead of adapting to customer needs, other buyers are sought for the same products.

IN THE MARKETING-ORIENTED BUSINESS:

- The focus is on the marketplace: customers, competitors and distribution
- Monitoring the market is a routine part of the business
- Change is recognised as inevitable and manageable
- Management is committed to strategic business and marketing planning and creative product planning
- The emphasis is on profit – not just volume, with profit and growth kept in balance.

Table 2.1 Production v. Market-oriented management

For example, a market-led business would never pursue production standards that delivered a level of finish on livestock which buyers see as unwanted fat. Many livestock producers nevertheless resisted the evidence of market signals in pursuit of their own idea of a finished animal. A market-led business would not have ignored customer demand for product assurance until it became a necessary condition for market access. A market-led business would not seek to cut costs by reducing stocks of marketable products to a level which threatened its ability to supply orders promptly or meet an unexpected surge in demand, since this could drive customers to alternative suppliers. Similarly, it would not judge its performance by production efficiency, technical excellence or financial management, but by customer satisfaction and competitiveness as measured by market share and revenue – which of course reflect production, technical, and financial performance.

The relevance of a management approach which puts customers and competitiveness at the heart of business planning should be obvious as farmers are increasingly exposed to market forces. At the very least it will throw a penetratingly critical light on current business management and performance, identifying unseen slack and opportunities for adding value. At the other extreme it may prompt a radical reappraisal which redirects management effort and investment into completely new enterprises.

The need for a re-appraisal of the business is urgent on many farms as the policy and planning framework changes. In the past, farm business planning was a relatively straightforward exercise, which concentrated on production targets and financial performance set by reference to ready-made farm management blueprints and industry profitability and performance standards. The exercise was also relatively risk-free in regulated markets, since sales could be forecast by reference to yields per hectare or per animal, and revenue to intervention prices and livestock premia etc. Today, the free information and advice which farmers once took for granted has to be bought in, adding a cost which many small businesses find difficult to afford.

This puts a premium on the ability to obtain and interpret information without outside management and marketing advice. Farmers will also have to rely increasingly on their own judgement in assessing enterprise viability, in setting production and financial targets, and in monitoring performance and fine-tuning the system. All this is the job of the marketing planning process, which can be built into the management routine without adding substantially to the management burden or requiring radically new skills: both vital in the owner-managed business – especially family farms – where the limiting resource is invariably labour.

THE MARKETING PLANNING PROCESS

This book takes the reader through the marketing planning process as a way of introducing marketing concepts and techniques relevant to the owner-managed business, and marketing terms which are in normal business use.

Like any other technical discipline marketing has its own terminology, which has a strong American input because marketing practice and the study of marketing both developed in the USA. Over the years other cultures have added their own terms where the American usage seemed inappropriate, and the result is an infinitely expanding vocabulary which can easily intimidate or

WHAT MARKETING MEANS

- Clarify business objectives
- Identify current strengths and weaknesses
- Clarify marketing strategies and objectives
- Identify marketing opportunities
- Identify environmental constraints
- Devise marketing programme
- Implement and control
- Measure and review
- Modify and develop

Figure 2.2 The marketing planning process

irritate the uninitiated. Behind the vocabulary lie some familiar ideas, however, which leads the sceptic to conclude that the terminology is only window dressing, and marketing is simply common sense.

Neither of these is true. At the level at which most small businessmen need to understand it, marketing certainly has a large slice of common sense, but there is ample evidence that managers fail to transform common sense into a viable business plan and effective management. Where they invariably need help, moreover, is where common sense does not supply the answers. An understanding of formal marketing planning criteria and procedures is therefore essential for business managers as well as for intending marketing professionals. The terminology must also be mastered, if only to avoid intimidation by marketing professionals and bank managers.

The marketing planning process is illustrated in Figure 2.2 as a series of consecutive stages. In practice it is a continuous process of planning, implementation and review in which information is fed back into the business, modifying initial assumptions and entailing management adjustments. Objectives are revised in the light of new practice or environmental change; new opportunities are identified which modify both objectives and management. The process is the same whether the business is a large or a small one, a single- or a multi-product enterprise. The small owner-managed business may have an advantage over the large multi-departmental business because production, marketing and financial management are all in the same hands, with no management structures and consultative procedures to obstruct the flow of information and adaptive response.

Like conventional farm management planning, marketing planning has a *strategic* dimension in relation to which *tactical* production and marketing decisions are made. The object of marketing strategies is to achieve the best possible match between resources and market opportunities, and ideally to establish a long-term competitive advantage by securing a customer base loyal to the product and the supplier.

Four basic marketing strategies exist which involve different levels of innovation and risk: market penetration, market development, product development, and diversification. The appropriateness of these to a product or a business must be evaluated in every situation (and is discussed below) in relation to:

- internal business objectives, resources and capability (Chapter 3)
- the *marketing environment* (external marketing conditions: Chapters 4-6)

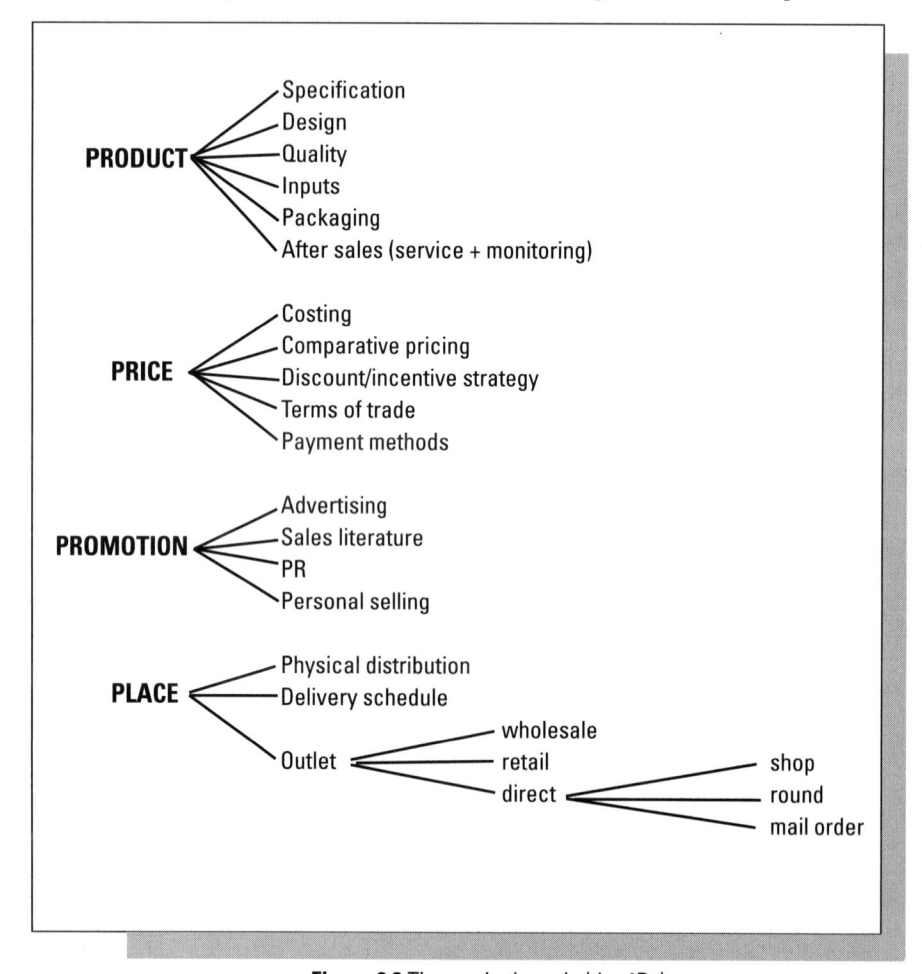

PRODUCT
- Specification
- Design
- Quality
- Inputs
- Packaging
- After sales (service + monitoring)

PRICE
- Costing
- Comparative pricing
- Discount/incentive strategy
- Terms of trade
- Payment methods

PROMOTION
- Advertising
- Sales literature
- PR
- Personal selling

PLACE
- Physical distribution
- Delivery schedule
- Outlet
  - wholesale
  - retail
  - direct
    - shop
    - round
    - mail order

Figure 2.3 The marketing mix (the 4Ps)

The outcome of this analysis is a sounder appreciation of the commitment involved in what farmers know as *niche marketing* and marketing professionals call *market segmentation*, where production is modified to supply a particular target market.

Niche marketing requires that the target market be identified and described through *market research*, and a *marketing programme* designed which specifies in detail all the production and marketing factors a manager can modify in order to deliver its requirements. This involves decisions in four main areas – invariably known as 'the 4 Ps' – which constitute what is known as the *marketing mix* (Figure 2.3):

- product (Chapter 7)
- price (Chapter 8)
- place (distribution – Chapters 9-10)
- promotion (Chapter 11).

All this information is brought together in the *marketing plan*, which has the same planning, implementation and control functions as a farm plan, directed now by marketing rather than production objectives (though production factors of course modify and are incorporated in the marketing plan). It is the marketing plan which also provides the all-important revenue estimates for the financial appraisal by determining costs in relation to a realistic sales projection – a difficult exercise when neither sales nor prices are guaranteed. The marketing plan is thus the basis for the enterprise feasibility and financial appraisal.

MARKETING INFORMATION

All this requires a substantial amount of information about the business, its markets, and the marketing environment. In the same way that farm records and technical information underpin good farm management, good marketing depends on reliable market and marketing information. Without this it is impossible to make rational choices about target markets, marketing strategy, product development, pricing formulas, the timing of the marketing effort, and so on. *Marketing research* – the collection, interpretation and use of information – is thus a key management function.

In the well-run business, up-to-date internal performance and profitability data will already be available, and the only new commitment is to obtain and store information about markets and the marketing environment: consumer trends, retail developments, advertised terms of trade, technical data, competitors' initiatives, government policy etc. This information is available from a variety of sources. Government and other official publications are a rich source, together with trade and marketing magazines, supplemented now by relatively inexpensive on-line information services accessible via the Internet. Other data sources which should not be overlooked include feedback from professional buyers and other farmers, public meetings, government agencies, farming unions, local authorities, regional development agencies and business support services.

There is in other words no shortage of information, but information is useless unless it is analysed. Raw data are precisely that: raw data which need interpretation to become useful intelligence. Consumer information, for instance, needs translating into a precise production and marketing schedule: exactly which product characteristics are needed? in which outlets? when? in what volumes? at what price? what promotion is required? and (most critically) can this be delivered at an acceptable profit? The information must also be easily retrievable. A basic recording system for marketing information is therefore essential in even the smallest business. Ideally this should be integrated with other farm records in a single computer data base which will facilitate retrieval and analysis, and reduce the need for and the cost of outside advice.

The information requirement is thus not a problem: what is difficult is the interpretation and application of information to the individual business situation. It is particularly difficult for an owner-manager to make objective judgements about internal business capability. Significantly, textbooks often refer to the process as a marketing *audit* – a term familiar from accounting, where objectivity is guaranteed by employing independent auditors.

Management and marketing consultants can of course supply a similar independent input at a price, and where management time is genuinely at a premium, or the manager cannot achieve the necessary objectivity, this may be a very worthwhile cost. Many owner-managers think they cannot afford this cost, or are sceptical about its worth even though they acknowledge that they have insufficient time to do a proper job of the appraisal. However, the costs of a failed project will almost certainly exceed those of expert advice. Although this book encourages a strong measure of self-reliance, it is therefore essential to recognise when outside help is needed. There are also some marketing activities where a professional input is almost invariably essential: for example, some market research, promotion, or product and packaging design. When an outside input is required, an understanding of marketing principles and techniques will nevertheless limit the input and its cost.

CONCLUSION

For many farmers, staying profitably in business may not require radically new departures which require new skills, but a reappraisal and reorganisation of the business will almost certainly be necessary to improve the capacity to respond to marketing opportunities. This can be achieved by a management approach which puts customer satisfaction and competitiveness at the heart of business planning and performance, which does not entail an unacceptable extra management burden. Without this there is serious risk of a misdirection of resources, and a probable decrease in revenue from the market which will increase dependency on income supports of one kind or another.

CHAPTER 3

Analysing the current business

All businesses would prefer to continue doing what they are already doing, using existing resources and well-tried management systems. If their traditional markets decline, however, or the business environment changes, they must adapt their skills and resources to supply a new demand. Even if a business is currently well-directed and profitability is acceptable, a sustainable business depends on identifying what constitutes profitability over time. The sensible manager is therefore always on the look-out for more profitable ways of marketing or adding value to existing output, or customers who would pay more or offer better terms.

Any new opportunities must be evaluated in relation to internal business capability, external constraints which may limit this capability, and the market requirements that must be supplied. The procedure is invariably known as a SWOT analysis, because it identifies business *strengths* and *weaknesses* relative to external *opportunities* and *threats* (Figure 3.1). The more radical the departure from the core business, the more thorough this analysis must be. Even a marginal change requires systematic appraisal, however, since any tendency to over-optimism or bias is almost certain to be penalised in the marketplace. The following four chapters indicate the kind of research and analysis that is required.

THE BUSINESS

Strengths
Long-established
Quality product
Strong service capability
Innovative capacity

Weaknesses
Product-oriented
Labour shortage
Few financial resources
No marketing research

THE MARKETING ENVIRONMENT

Opportunities
Export markets
New market segments
Loss of competitor
Promotional campaign

Threats
Competition
New legislation
Economic recession
World trade situation

Figure 3.1 SWOT analysis

BUSINESS CAPABILITY APPRAISAL

In the well-run business, physical and financial information about the business will be available which will simplify the marketing audit and reduce the need for outside advice – though one outcome of the audit may be to indicate the need for an outside input. The complexity of the analysis will reflect that of the business, but the same questions will always need answering:

Current marketing performance
- What business are we in?
- What are the business objectives?
- What resources do we have or could we reasonably acquire?
- How good are we at what we do?

Marketing strategy
- Where do we want to go?
- What degree of market segmentation is appropriate?
- What do we have to do to beat the competition?
- Can we do it profitably?

WHAT BUSINESS ARE WE IN?

'What business are we in?' may seem self-evident, but it is the first question an outside consultant would ask. Most farmers would reply that they are in sheep, cereals, or milk etc., although their business is in fact the profitable utilisation of a land resource and management expertise, which simply happen to be employed in sheep, cereals, or milk production.

This is not a hair-splitting definition, but a necessary warning against 'product myopia': a short-sighted preoccupation with existing output which results in missed opportunities and unperceived signs that a business is out of touch with its market. The example often quoted is that of the American railroad company which lost its business to alternative transportation systems because it failed to see that it was in passenger and freight transport, not railroads. Many farmers suffer from the same product myopia in not acknowledging that their business is the productive use of land and other resources, not the production of long-established commodities. Seen in this light the business may be far from optimally directed, and acknowledging this is the first step in shifting the management focus from a production to a marketing orientation. It may also become essential if income from food production has to be supplemented or replaced by revenue from other land-based goods and services.

WHAT ARE THE BUSINESS OBJECTIVES?

In a large company, a statement of corporate objectives is clearly necessary to ensure effective coordination between different operating divisions, and to prevent the pursuit of incompatible objectives: for example, production efficiency, financial security, market growth. A statement of business objectives is also normal practice in a farming partnership or a company with non-executive shareholders or outside investors. In the owner-managed business, by contrast, a formal statement of business objectives is often overlooked, although failure to clarify the business objectives is a common reason for under-achievement and misdirection of resources. Typically, different members of a family business – operating divisions – pursue different goals, or the owner-manager does not see the incompatibility of his own goals.

A clear statement of objectives is therefore essential in even the smallest business, and to be useful, it must meet several requirements which are remembered by the acronym SMART:

- *Specific*: identifying a stated percentage of market share, turnover, acceptable throughput, occupancy level of leisure facility, accommodation etc.
- *Measurable*: sales volume, profit levels, employment generated etc.
- *Achievable*: the necessary resources must be available or procurable; the enterprise of appropriate scale
- *Realistic*: the enterprise should reflect the resource base, and allow for compromise (say) between short-term profit goals and long-term survival in changed environmental conditions
- *Time-based*: staged/scheduled implementation leading to measurable achievement.

The value of such a statement of objectives will become particularly evident when unexpected events threaten to divert management from its long-term course, sending it into a cul-de-sac from which it is later difficult to extricate the business. External circumstances will still intervene, forcing all managers to make decisions on the basis of imperfect information, but lurching from one short-term decision to another without a clear sense of overall purpose is a recipe for disaster.

To avoid potential conflict within the business it is important to acknowledge *all* the business objectives (Table 3.1) and their relative priorities. For example, although economists assume profit maximisation to be the reason for being in business, it is rarely the sole objective of many entrepreneurs or self-employed people, even though their profit is their income. In practice profit maximisation is qualified by other objectives like job satisfaction, or sustaining a family business for the next generation. Business diversification in particular is often motivated not by profit expectations, but by the wish to spread the business risk or employ an unemployed family member or an unused skill.

Performance objectives are not difficult to identify (total output, return on capital/assets, profitability etc.). Farm businesses are also used to setting internal objectives (or at least measuring efficiency) in yield per hectare or per animal, and these efficiency measures are routinely compared with standards for the industry or for a high-performing sector (say, the best 10 or 20 per cent).

Some *external objectives* are easily identified: for example, financial targets to satisfy a bank or other lender, or planning requirements which entail production modifications and additional costs. Other external objectives may never have been written into the business plan although they affect day-to-day management and the achievement of long-run business targets: for instance, meeting public expectations on animal welfare, access, or environmental management, or not offending local opinion by insensitive planning applications and nuisance from farming activity.

Directional objectives	
Market leadership, measured by:	competitive position degree of innovation technological advances
Market spread, measured by:	number of product markets number of customer segments number of countries
Customer service, measured by:	product utility product quality product reliability product availability
Performance objectives	
Growth, measured by:	sales revenue volume output market share profit margin contribution
Profitability, measured by:	return on capital employed return on assets profit margin on sales revenue
Customer satisfaction, measured by:	repeat orders logistic benchmarking goods returned complaints lost customers
Internal objectives	
Efficiency, measured by:	sales on total assets stock turnover credit period liquidity department costs on sales
Personnel, measured by:	employee relations and morale personal development average employee remuneration sales revenue per employee
External objectives	
Social responsibility, measured by:	corporate image price/profit relationship resource utilisation public activity community welfare ethical standards

Table 3.1 Business Objectives

Like most businessmen, farmers would probably not recognise the term *directional objectives*, but typical of these would be a wish to have a range of buyers in order to achieve a degree of marketing flexibility and independence. This may be particularly important for a raw material supplier, even though it runs against the industry trend towards specialisation and servicing of one or a few outlets, and may present operational problems as well as a difficult strategic choice. Customer service and product quality are other directional objectives which may never have been explicitly stated, although they have a high priority for most farm-based businesses and should certainly figure prominently in any marketing programme. Other directional objectives which may not be acknowledged, although they are expressed in many farmers' management style, include the wish to be seen as highly innovative or technologically advanced, or to be socially responsible.

From these examples it is clear that business objectives are frequently incompatible. To prevent conflict, trade-offs may therefore be required which modify operational and performance targets. The primary objective of an organic farmer, for example, may be food production for a living, but his organic philosophy (a directional objective) will have implications for the marketing mix as well as for production. He may disapprove of supermarkets and wish to sell direct to consumers or via small family shops, but he may also nurture an ambition to become the largest organic producer in the country or head a large marketing operation. These divergent objectives must therefore be acknowledged and reconciled.

WHAT RESOURCES ARE AVAILABLE?

In seeking to achieve its objectives management deploys physical, financial and human resources. An established business will have some resources in all three categories, but more will generally be needed in order to meet reorganisation or development targets. In both cases it is important to identify any currently unused resources with development potential – which may include human skills as well as neglected physical resources like watercourses, woods, redundant buildings etc. It is also important to identify *under*-utilised resources which might be redeployed, in order to reduce the need for an additional input. However, these must be evaluated for their fitness for the job, and a sale value or redundancy cost must be assigned to them as an indication of the true cost of redeploying them.

In a well-managed business the opportunities for redeployment should be minimal, so additional resources will invariably be required to support expansion or development. If a change of direction is intended, existing resources which may become redundant could have a useful salvage value, but it may be insufficient to fund the new venture. This may force the abandonment of the proposed change – a common situation in all industries, and the main reason why radical changes of business direction are relatively rare.

In practice, finance for a well-researched and well-presented change can generally be obtained; a more serious obstacle, commonly underestimated, is inadequate or inappropriate management expertise. This can be overcome, but finding the time to develop new skills may be impossible in the small owner-managed business. An inappropriate business 'culture' is more difficult to overcome, and frequently becomes evident only after a new enterprise comes on stream. Typically, entrepreneurs fail to anticipate and adapt to marketing discipline, particularly where it involves routine contact with the customer. Their personal commitment to the new venture may also prove insufficient once the full management implications become clear. Options closer to established management competence may therefore be more appropriate than striking out in a totally new direction, and may achieve the same objectives.

Once the available and required resources for a new initiative have been identified, the feasibility of procuring additional resources can be determined, together with their cost. At this preliminary stage the precise amount of the additional input will be unknown, and the necessary information may not become available until production and marketing plans are well advanced. It should nevertheless be possible to ascertain from the current resource base whether the additional resources are likely to be acquired. The current level of net worth will be important in securing finance for additional land, buildings and plant, and for the redevelopment of existing facilities. Current asset structure may also be an important consideration for lending institutions, though they are increasingly more interested in business performance, an assessment of which will normally be expected to accompany the business accounts.

HOW GOOD ARE WE AT WHAT WE DO?

There is no shortage of technical and financial measures against which farm enterprise performance can be measured, and most farmers will already have an idea of their comparative position by reference to published industry and sector standards. In a single-product business, weakness may already be apparent in falling revenue, so the possibility of earning a premium by production or marketing changes needs urgent investigation. In the multi-product business, most managers will be aware of the relative income contribution of different products; they may nevertheless have difficulty in deciding where management effort and resources should be concentrated in order to maximise their effect, given that these resources are invariably limited.

The resource allocation decision is further complicated in a diversified farm business with alternative enterprises, for which few industry or sector measures of best practice exist. The industry/sector standards used for farm management appraisal are based on data collected from representative samples of farm businesses, systematically analysed and published by governments. No similar data base exists outside agriculture, where business confidentiality limits the availability of information about competitors. More data is becoming available as governments encourage firms in all sectors to adopt *benchmarking* – a new term for the comparative business performance analysis which agriculture has always practised. For the foreseeable future, however, managers of alternative enterprises will have to set their own performance targets in the absence of published information about industry or sector best practice.

The kind of data which is required is summarised in Figure 3.2. Familiar quantitative measures form part of the analysis, and these may be compared with any available published sector information: for example, sales achieved/sales forecasts, sales costs, market share growth, financial performance (gross profit, net profit, return on investment). This information must be supplemented by qualitative analysis of the requirements of a given market (for example, price v. convenience, or different quality requirements), and the degree to which current output satisfies them.

Such data is obtained through market research, which is discussed more fully in Chapter 6. For an indication of its relevance to the capability appraisal under discussion here readers are referred to Box 3.1, which reproduces a market research questionnaire recently used by a farmers' supply cooperative to 'benchmark' its performance in comparison with direct-supply competitors. The output from the questionnaire indicated a reassuringly high level of customer satisfaction with the cooperative's performance, but identified one or two areas where direct-supply competitors had a slight advantage, and others in which the cooperative had the competitive edge. This information allowed management to concentrate attention and resources on the weaknesses identified in its current offer, while revising its long-run strategy to build on identified competitive strengths.

Market research techniques are particularly useful to the owner manager because they help to provide the objectivity necessary to acknowledge internal weaknesses, and to identify effective ways of remedying or circumventing them. Market research is thus not the exclusive province of big business, but a valuable marketing tool for the small, owner-managed business.

How good are we at what we do?

Data required:

Product/service portfolio
Existing customers
Potential customers
Customer profiles
Existing cost/budget/profit levels
Current marketing activities
Sales and promotion effort
Business ('corporate') image
Competitor offers

Figure 3.2 How good are we at what we do?

PRODUCT/BUSINESS PORTFOLIO ANALYSIS

Another valuable tool for the owner manager is *product* or *business portfolio analysis* – a formidable term for a technique which identifies the relative contribution to the business of existing products or enterprises, to provide an objective basis for resource allocation and management decisions.

Several techniques of product portfolio analysis exist, but the one most commonly used in all business sectors is the Boston Box, named after the Boston Consulting Group which devised it. This locates products and enterprises in a matrix representing their relative revenue contribution to the business, together with their market share and growth prospects (Figure 3.3). Four categories of product/enterprise are identified:

- a *dog* has a low market share and low or declining growth
- a *cash cow* has a high market share in a slowly growing market
- a *star* has a high market share and high growth prospects
- a *question mark* has a low market share but high growth prospects.

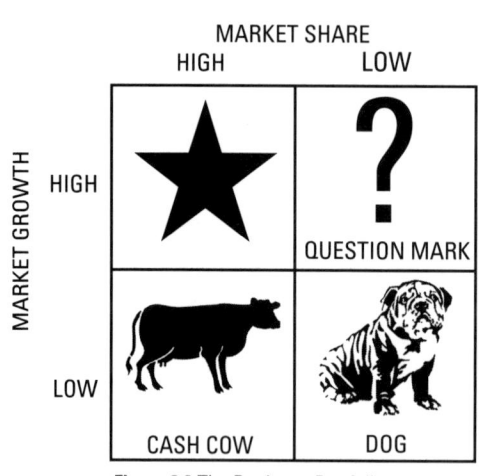

Figure 3.3 The Business Portfolio

No individual farmer or small businessman marketing his own, generally unbranded, product is likely to have high market share products. What he may have are products in sectors with a high market share (for instance, milk, sugar beet, potatoes) which represent a high proportion of his revenue. He cannot afford to allow these cash cows to fall behind technologically or to become outclassed in the marketplace; consequently, their management and revenue capacity must never be sacrificed or neglected in the pursuit of new marketing opportunities and enterprises.

The revenue from cash cow products and enterprises is available for investment elsewhere in the business, possibly in question mark products which it may be possible to move into the star category (high growth and high share). This invariably means upgrading the product and its management – a process which does not stop once star status has been achieved, since potential stars need to be treated like cash cows if they are to maintain their revenue potential. This requires continuous upgrading and repositioning of the product to maximise its revenue potential over a long product life cycle (Chapter 7), which entails a continuing financial investment. If the strategy succeeds there will be a high pay off, but if the strategy fails, question mark products may become dogs. An interesting example is organic food, which is generally believed to be a potential star enterprise, but could also be a dog if predicted growth fails to materialise.

In the case of dog products the situation is less clear. It may appear sensible to abandon such poor performers, and concentrate attention and resources on the rest of the product portfolio. However, this may be short-sighted, since a dog may simply be under-performing because it has been neglected; a redirection of management attention and limited resources could therefore make a major improvement. Some dog products may also have an important effect on the overall business profile, and their deletion from the product portfolio may undermine customer satisfaction with other products, or customer perception of the business as a whole. For example, a leisure facility attached to farmhouse accommodation may not itself generate revenue, but its absence could reduce the attractiveness and financial performance of the accommodation enterprise. Similarly, a pick-your-own facility alongside a farm shop frequently does not make money, but attracts visitors who buy other products and services.

Product width and depth

Individual product decisions must therefore be made in relation to the overall credibility of the product portfolio, and they must also take into account the width and depth of the product range. Product *width* is the term used for the

> **A customer service questionnaire used by a farmers' supply cooperative to evaluate its service performance**
>
> How would you rate our service on the following scale?
> (1 = very poor, 2 = poor, 3 = good, 4 = very good, 5 = excellent)
>
> *please circle*
>
> 1. Frequency of delivery — 1 2 3 4 5
> 2. Time from order to delivery — 1 2 3 4 5
> 3. Reliability of delivery — 1 2 3 4 5
> 4. Emergency deliveries when required — 1 2 3 4 5
> 5. Stock availability and continuity of supply — 1 2 3 4 5
> 6. Orders filled completely — 1 2 3 4 5
> 7. Advice on non-availability — 1 2 3 4 5
> 8. Convenience of placing order — 1 2 3 4 5
> 9. Acknowledgement of order — 1 2 3 4 5
> 10. Accuracy of invoices — 1 2 3 4 5
> 11. Quality of sales representation — 1 2 3 4 5
> 12. Regular calls by sales reps — 1 2 3 4 5
> 13. In-store merchandising support — 1 2 3 4 5
> 14. Credit terms offered — 1 2 3 4 5
> 15. Customer query handling — 1 2 3 4 5
> 16. Quality of outer packaging — 1 2 3 4 5
> 17. Well-stacked packaging — 1 2 3 4 5
> 18. Easy-to-read use-by date — 1 2 3 4 5
> 19. In-store quality handling and display — 1 2 3 4 5
> 20. Friendliness of staff — 1 2 3 4 5

Box 3.1

number of products in a business offer; product *depth* is the variety within each product 'line'. For example, a farm shop may sell fresh produce, grocery products, frozen goods, and bakery (product width), and within each product category there may be varying product depth (Figure 3.4).

Reducing the width of the product range by the deletion of a complete product line or enterprise may have dramatic effects on resource use and profitability: for example, going completely out of potatoes, or deleting frozen goods in a farm shop. Reducing the depth of the product portfolio (by growing only one variety of potatoes, or selling ice-cream but no other frozen products) will generally have less dramatic effects, but may still be substantial enough to have a significant knock-on effect on the business. Customer expectations of the product and their reasons for trading with the business must therefore be identified *before* deleting a product whose individual financial contribution is currently unsatisfactory, but nonetheless contributes to the business offer and total revenue.

The consistency of the product range is a further consideration, since it too may affect customer perceptions of the business. Doorstep milk delivery is one example where the introduction of some goods into the product mix was seen as inappropriate by some customers: for instance, potatoes were considered 'dirty' alongside 'clean' milk. A UK multiple food retailer who introduced car sales alongside food stores failed to attract buyers because they did not associate food supermarkets with the purchase of a car. A 'noisy' facility alongside farmhouse accommodation may similarly be inconsistent with customer perceptions of the product, which offers a peaceful countryside retreat.

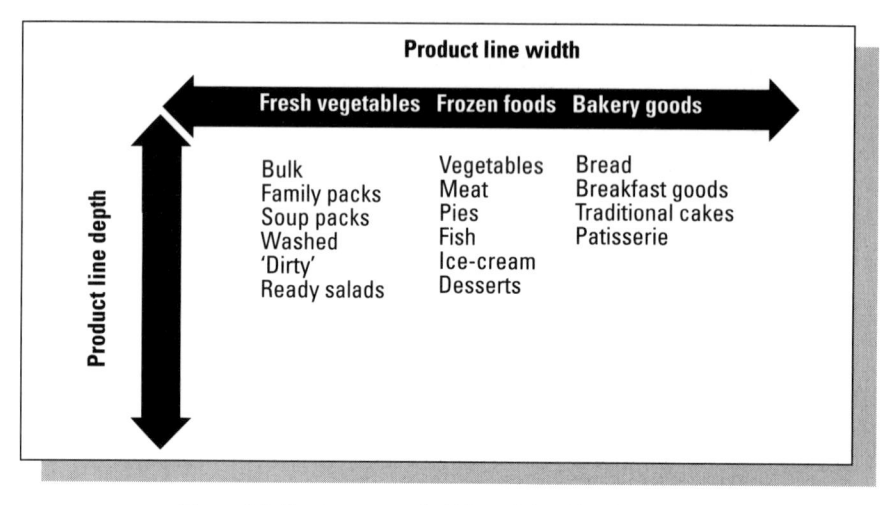

Figure 3.4 Product range (width and depth): a farm shop

WHAT BUSINESS COULD WE BE IN?

All the information about a business and its current market can be brought together in a matrix which will indicate any areas where marketing performance may need improvement (Figure 3.5). In a situation of low business strength and poor market prospects the right strategy may be to abandon the product or the business completely – an extreme position which may have to be confronted if other strategies are not viable. At the other extreme, a situation of strong business strength and strong market prospects may call for no change, but careful monitoring is necessary to identify any sudden change in customer requirements or environmental factors.

Most businesses fall somewhere between these two extremes, in a position where a change of strategy is indicated, but unclear. A strong business with poor market prospects, for instance, could improve its current products, diversify its range, or attempt to reposition itself in the marketplace. In pursuing any of these strategies it would enjoy the advantage of underlying strength, which may allow it to find the management and financial resources to improve its market attractiveness. A weak business with a high level of attractiveness may be a take-over prospect for other firms, as some diversified farm enterprises have discovered. Alternatively, it may sell off or otherwise capitalise on one of its products in order to acquire resources for redeployment and restructuring (by selling milk quota, for instance, or by franchising cheese manufacture).

		Poor	Average	Strong
Business Capability	Poor	ABANDON ENTERPRISE	PHASED WITHDRAWAL	TAKEOVER TARGET DOUBLE OR QUIT
	Average	PHASED WITHDRAWAL		TRY HARDER
	High	IMPROVE DIVERSIFY REPOSITION	GROW	LEADER NO CHANGE BUT MONITOR

Figure 3.5 Market prospects

MARKETING STRATEGIES

On the basis of this internal capability appraisal it is possible to make rational choices between alternative marketing strategies. Four broad marketing strategies are available to any business: market penetration, product development, market development, and diversification. All four strategies require information about existing and potential customers, and the product attributes necessary to gain some competitive advantage. They require different business strengths and different kinds of marketing commitment, however, and have different implications for the marketing mix.

- *Market penetration* seeks to increase market share for existing products, and usually requires more promotional activity and efforts to extend the scope of distribution. In a saturated market, success depends on taking market share from the competition.

- *Product development* relies on knowledge of existing markets to introduce modified or new products. It requires production skills to give products a competitive edge, and is normally a long-run strategy because it takes time to develop and re-position products. It may be a higher-cost strategy, but has the advantage of operating in known markets.

- *Market development* seeks to move into new markets with the same products. This puts the management emphasis on market research, to identify and move into new market segments or geographical areas – for example, export markets. This is a higher-cost strategy than market penetration, but is usually cheaper than product development.

- *Diversification* entails the introduction of new products into new markets, usually by moving into completely new lines of business alongside established enterprises. In most business sectors this is the highest-cost and the highest-risk strategy, and entails the highest production and marketing commitment.

Market penetration and product development are in other words *product-based* strategies which exploit production strength, and farmers have routinely adopted both strategies in expanding production of existing commodities and introducing new crops, varieties, livestock breeds, and up-graded technology. Market development and diversification are *market-led* strategies which require marketing skills rather than production strength; it is therefore not surprising that most farmers were slow to adopt either strategy as long as production was unrestricted and commodity prices were subsidised.

The risk associated with these strategies is normally assumed to increase as the business moves away from its core enterprise. In theory, the lowest-risk strategy is always to start from what is known. Most firms – and most farmers – consequently think first of market penetration (expanding output) and secondly of product development, preferring to build on existing strengths with an established product and established markets. Businesses at the limit of their production capacity in existing products and markets must sooner or later consider market development. Diversification is usually described as the last strategy to be considered because of its greater risk.

In fact, diversification was the first strategy to be adopted by many farming families because it exploited unused or under-used resources. Market development, by contrast, is seen by farmers as the highest-risk strategy because it is the greatest unknown. Market development is also commonly assumed to be the province of large companies which employ professional marketing staff and can call on large budgets to create new markets. Small businesses are actually very good at market-led strategies, however, because they tend to be close to their customers. Many – particularly in rural areas – also have a limited customer base, and to retain existing customers and attract new ones they are obliged continually to introduce new products and move into new product areas: a situation classically illustrated by farm shops. Market development work by producer organisations has also been extremely effective in several farming sectors and rural enterprise, notably export markets and farm tourism.

Any or all of these marketing strategies may be employed simultaneously, provided the resource base and management capacity are not over-stretched in the process. For example, dairy farmers who were prevented from expanding milk output *diversified* into ice-cream manufacture, quickly added new flavours (*product development*) in order to increase *market penetration*, and also developed *new markets* (caterers, a wider geographical area).

MARKET SEGMENTATION

The implementation of any of these marketing strategies depends on a further strategic decision:

- whether to address the total market for a product with a single marketing mix, *or*
- whether to research and identify market segments which may be supplied with a different product and marketing mix to match their separate needs.

The former is known as *undifferentiated*, *commodity* or *mass* marketing; the latter is what farmers know as niche marketing, and textbooks refer to as *differentiated* marketing or *market segmentation*.

Undifferentiated marketing

Undifferentiated marketing assumes that demand is *homogeneous*: that the needs of individual customers are similar enough to be satisfied by a single marketing mix, requiring no product variation, a single price regime, one distribution system, and unified promotion. The classic example is the Ford Model-T, which supplied the needs of a new mass market by providing a highly standardised product 'in any colour as long as it was black'.

As this example suggests, this take-it-or-leave-it approach is most likely to be effective where demand exceeds supply (when there is a highly novel product and/or an absolute supply deficit). It may also be effective in other situations where customers really do have similar requirements. It is thus likely to be most appropriate for products with very little or no intrinsic variation, for which demand is also relatively homogeneous (petrol, basic food staples), or about which customers do not care sufficiently to distinguish between them. As a producer strategy it has the obvious advantage of limiting the marketing commitment, but it tends to rely on competitive pricing, with the object of maximising sales. Profit therefore depends on high sales volume and low costs, and the strategy is usually associated with the drive for economies of scale.

> *"There is no such thing as a commodity - it's what you do with it! Change the specification, redesign the label and alter the packaging."*
>
> **Managing Director, Dairy Crest Liquid Products**
>
> **Box 3.2**

Mass marketing is characteristic of agricultural commodity markets (hence the term 'commodity marketing'), because there is little producer incentive to target customers in subsidised markets. However, it is not characteristic of consumer food markets, which are characterised by oversupply, world-wide competition for sales, and greatly increased consumer discrimination and product choice. Even the most basic staple food purchases are now differentiated at the consumer level: retail milk by fat content, pack size, and even breed of cow; fresh potatoes by variety, pack size, end use or organic source. At the industrial level this is evident in exacting specifications as buyers seek to ensure traceability and greater product consistency, and suppliers compete to supply them by establishing the identity and superiority of their product over others in the marketplace.

Differentiated marketing

Differentiated marketing recognises that demand is typically *heterogeneous*: composed of distinguishable sub-markets *(segments)* whose requirements can be described sufficiently to underpin a special marketing mix and product offer. A business which can 'segment' the market may therefore be able to gain a competitive advantage by supplying a particular demand: for example, organic vegetables, grass-fed beef, outdoor pork, out-of-season milk of given compositional character.

The degree of segmentation practised may vary widely. At one extreme an existing product or marketing method may require only marginal modifications to allow a given demand to be targeted. This strategy is referred to as product differentiation, which seeks to maximise revenue from existing output. At the other extreme, a business may set out to identify from scratch an unsupplied demand which it may decide to target with a 'designer product' tailored exactly to meet its requirements: this is market segmentation as the purists know it.

In many businesses (certainly in many farm businesses), maximising the potential of existing products and enterprises is the only realistic option, within a broad market penetration or market development strategy. Since this strategy does not produce a product materially different from competing offers, it tends to rely heavily on product branding and promotion, and/or price-related incentives to establish customer loyalty for an existing product. However, establishing a brand is difficult, particularly in a highly competitive sector characterised by strong manufacturer and retailer brands (Chapter 5). The promotional effort necessary to maintain customer awareness of a brand also tends to be very expensive, and is difficult to achieve for agricultural raw materials.

Competitive pricing is certainly an obvious way to differentiate a product, since a rational purchaser of unbranded commodities (basic products like milk, petrol, or common industrial inputs) will generally buy the cheapest. Lower prices may therefore be one way to penetrate markets. However, as a long-term strategy this is only sustainable by businesses with low costs, achieved through efficient use of resources (often through the exploitation of economies of scale) or by access to low-cost resources. A price-led product differentiation strategy may also lead to a slippery slope of downward prices which is only too familiar to producers of industrial inputs.

THE SEGMENTATION DECISION

The combination of these two elements – differentiation and production costs – indicates four potential segmentation strategies as shown in Figure 3.6. Two of the strategies are self-evident. Any company which could achieve a high degree of product differentiation at a very low price ought to be outstandingly successful. At the other extreme, high costs combined with poor product differentiation are likely to be unsuccessful, for although everyone would like a Rolls Royce for the price of a Ford, no-one will pay Rolls Royce prices for a Ford.

Most businesses will again fall between the two extremes. Some seek to serve the whole market with a standard product at a relatively low price achieved by low costs. This 'cost leadership' is seen in the agrifood sector when farmers sell feedwheat, and some food manufacturers and retailers sell basic products like milk and cheese, on the basis of price. Other businesses seek deliberately to identify niche markets whose distinguishable product requirements they can profitably supply with a specific marketing mix.

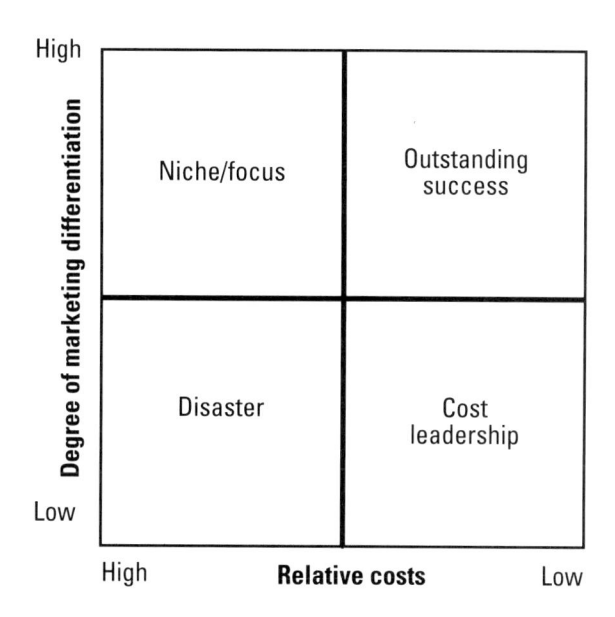

	Relative costs High	Relative costs Low
Degree of marketing differentiation High	Niche/focus	Outstanding success
Degree of marketing differentiation Low	Disaster	Cost leadership

Figure 3.6

The final choice will depend on the business strengths and weaknesses identified, the degree of management commitment and risk entailed, and the accuracy with which a segment and its requirements can be described (Chapter 6). In almost every business, however, it is both possible and desirable to make greater or smaller modifications to a product and/or its marketing in order to target a market which provides greater security, if not a marketing premium. Until recently, most farmers were interested in market segmentation only if a clear revenue advantage could be demonstrated in the form of a better price or more sales; the food scares of the 1990s which closed some markets completely to producers have nevertheless demonstrated the importance of secure market access.

On the negative side, market segmentation tends to incur higher marketing costs. Care is therefore needed in evaluating its financial feasibility, and strict cost control is required over marketing as well as production. Market segmentation also tends to involve greater risk, which increases as the segment is more finely drawn and the product more closely tailored to its specific needs. As one analogy puts it, market segmentation is like a rifle aimed very carefully at a small target: if you hit the target you will be very successful, but if the target moves even slightly or an external event intervenes, you could miss altogether. Common sense therefore suggests the wisdom of a multi-segment strategy, rather than a concentration strategy which risks all the eggs in one basket by supplying a single, vulnerable market segment.

Market segmentation is possible both in industrial and consumer markets. In practice there is little difference between the psychology and purchasing behaviour of consumers and industrial buyers (Chapter 6), and the rifle technique can be very successful at the level of the firsthand buyers with whom farmers primarily deal. In industrial markets the risk associated with market segmentation can moreover be reduced through contracts binding on both parties (Chapter 9). At the consumer level contractual purchase is generally possible only where the ultimate form of segmentation is practised: production to order (perhaps of one-off products to an individual customer's specification), where the risk is minimised by a prior purchase agreement. What is commoner in consumer and industrial markets is the production of a standard range of products which may be customised by marginal modifications to suit different customers.

For large producers who need large volumes in order to remain profitable this is a normal way of coping with the need to target products at different customer segments. The production of large volumes of identical products brings economies of scale from specialisation and the use of technology, and market segmentation takes effect at the post-production level, through added services. This strategy is particularly appropriate to the marketing of necessities and run-of-the-mill products to which different services can be attached for different buyers. The market segments identified may be strictly *product-based* (a lamb of given breeding, weight and conformation; milk of a given bacterial count and composition), and/or *service-related* (lambs in given numbers marketed at pre-arranged times; milk traceability, level delivery).

As readers will have recognised, market segmentation is therefore not a new phenomenon. Producers of most farm commodities have always targeted at least some of their output at a particular buyer in a given market, or at particular markets with a reputation for attracting certain types of buyer who require a particular product. Lamb producers, for example, recognise many market segments requiring different weight ranges, different levels of conformation and fat cover etc., and they adapt production and marketing to supply those in which they can achieve the targeted characteristic. However, market segmentation is likely to become an essential strategy for most producers as market support for mainstream commodities declines.

In pursuing market segmentation it is generally assumed that it is better to identify an unserved segment rather than to confront existing competition on its own ground. The fact that there is no present competitor in a market gap is no guarantee that a new offer will go unchallenged, however. An innovative offer is too easily copied by another business which has an established customer base or a production or marketing advantage. This makes it easier for the established competitor to win sales with a 'me-too offer', benefiting from the cost and effort which the innovator invested in developing the niche.

By setting up in competition with an existing business it may be possible to reverse this process by actively seeking to win some of its customers with a better offer. This strategy would also substantially reduce one of the major costs and difficulties for a new business: advertising and promotion. It is nevertheless a high-risk strategy, since confronting an established competitor head-on requires a very effective marketing mix to ensure that the new offer really is better than the competition. More detailed market research is therefore required, in order to identify exactly which product attributes must be supplied in order to hit the target.

CONCLUSION

In almost every business, market segmentation can and should be adopted for at least some of the output, since it is generally better to serve some customers well and retain their custom rather than supply all customers fairly well, and risk losing them to the niche marketer who serves them better. The ability to achieve this depends on an objective appraisal of internal business capability which this chapter has outlined, together with an equivalent understanding of the marketing environment and customer requirements which is outlined in the following chapters.

CHAPTER 4

Scanning the marketing environment

The ability to exploit marketing opportunities depends not just on internal business capability, but on the marketing environment: everything from competitors and suppliers to government policy and the state of the world economy (Figure 4.1). Although these factors are clearly beyond the manager's control, their potential impact on the business must be understood. More importantly, the most successful businesses in all sectors have been shown to be those which routinely scan the marketing environment for early warning signs of trouble which allow avoiding action, and market signals which identify opportunities ahead of the competition.

The terms *macro* and *micro* environment are widely used to distinguish between factors which are common to all businesses regardless of sector, and

Figure 4.1 The marketing environment

factors which are unique to every business. Macro-environmental factors normally include national and international politics, the economic situation, the legal framework and the natural environment, to which business plans and management must simply be accommodated. A firm's micro environment is generally defined as its customers, competitors, distributors and suppliers, which managers can allow for by adapting the marketing strategy and the marketing mix. An alternative name for the micro environment is hence 'task environment', because this is where the management task is assumed to lie.

In most farm businesses this distinction is not very helpful, though this chapter respects the division because the terms will be heard in normal business use. The experience of most farm businesses is that natural and political factors demand more management attention and can have a much greater impact on the business than customers, competitors and distributors; they therefore figure prominently in the farmer's 'task'. Climate and physical situation exercise a major impact on agricultural production, and they may also be critical in alternative farm enterprises: for instance, sales of many products such as ice-cream, farmhouse accommodation and outdoor leisure activities are weather-related. Government policy regulates many aspects of farm businesses, and occasionally demolishes the best-laid plans overnight. Memorable examples are the overnight imposition of milk quotas by the EU in 1984, and the BSE crisis and subsequent collapse of the beef market in 1996 – each the result of political action which farmers were powerless to change.

On the other hand, although the political and the natural environment may be beyond management control in the short term, compensating action may be easier to implement than it is in the case of customers, competitors, distributors and suppliers. Farming systems can be modified to reduce the impact of physical or climatic disadvantages (different crop varieties, livestock breeds, husbandry regimes, irrigation, winter housing etc.). Indoor facilities and adequate insurance cover will protect against weather-related loss of business to a leisure or accommodation enterprise. Efforts can also be made to change the long-term situation. Farmers and other trade organisations can lobby governments and bring business interests to bear on the legislative and political process; public opinion may be influenced by effective promotion and educational effort designed to correct damaging perceptions.

From the manager's point of view, the textbook distinction between macro- and micro-environmental factors is therefore less important than the ability to distinguish between:

- *factors which can be controlled directly via the marketing strategy and the marketing mix* – for example, a product mix or marketing strategy to counteract seasonal imbalances in demand; a mix of outlets which prevents over-dependence on any single distributor; routine investigation of likely as well as current technical and legislative requirements on food processing and handling

- *factors which cannot be controlled via the marketing strategy and the marketing mix*, but may be susceptible to other, long-term influence.

This will not eliminate the risk to the business of major unforeseen events – as the introduction of milk quotas and the BSE crisis demonstrated. However, it will develop a rapid response capability which allows a business to adapt promptly and effectively to crises, and may turn a crisis into an opportunity. For instance, the imposition of quotas was a short-term disaster which closed the traditional business expansion route to milk producers, but led to the introduction of added-value and alternative enterprises which transformed the profitability and viability of many farm businesses.

THE MACRO ENVIRONMENT

Four divisions of the macro environment are normally recognised – *political, economic, socio-cultural* and *technological*, summarised in the suggestive acronym PEST (Table 4.1).

Prominent in any farmer's list of 'pests' must be the political environment, for farm businesses are arguably unique in the extent to which they have been directed, and their revenue supported, by political intervention. In the 1990s this support policy has been seriously questioned, and the outcome is uncertain in many countries and regional blocs like the EU. Over the same period the impact of environment and rural development policies has grown, introducing new business constraints and providing new income opportunities which reflect new public priorities.

Agricultural and rural policy will consequently remain a major management preoccupation for farming and rural business, and the following outline provides a framework within which changing priorities and their management implications may be more readily understood.

AGRICULTURAL POLICY

Almost all countries in the world have agricultural and food policies which managers cannot control, except through the influence they can exert on the political process. The original motivation for these policies was to ensure a reliable food supply by supporting farm incomes. As supply increased the objective became the support of farm incomes as a means of supporting rural communities. The result was production surpluses and a financial burden which is now questioned, not least because it has not achieved the objective of supporting farm incomes or rural communities.

Agricultural policies are invariably very complex, with detailed rules which must be complied with because they have the power of law. For example, the EU Common Agricultural Policy (CAP) is embedded in Community law, and is the regulating framework of agricultural production and marketing within the bloc.

Agricultural policies worldwide address three issues which significantly affect a producer's marketing options: the quantity produced; the prices to be paid for products; production methods, and their effect on the product and the environment.

Political/legal factors

National, local government, regional blocs, international politics
Political ideologies, lobbies, pressure groups
Financial support
Statutory legislation
Codes of practice
Consumer protection
Sector-specific policy

Economic factors

National/international economic situation (recession/boom; inflation; unemployment; income levels, interest rates etc.)
Exchange rates

Socio-cultural factors

Demographics (e.g. age structure of population)
Geographics (customer location, market distribution etc.)
Socio-economic groups
Lifestyles, habits
Norms and values

Technological factors

Innovation
Technology transfer
Beneficial effects (product opportunities, added leisure time, living standards)
Adverse impact (health, environment, other products)

Table 4.1 The macro marketing environment

Quantity Produced

Governments anxious to ensure adequate food supplies have traditionally provided price incentives and funded research to expand production. Farmers responded so efficiently to this that most developed countries and trading groups have become much more than self-sufficient. Policy-makers consequently seek now to restrain output via a range of methods which includes product price reductions together with quantitative restrictions on inputs, on the marketable quantity, or the quantity to which a specific price applies.

The CAP uses all these forms of restriction, so the producer's most basic marketing decision – what and how much to produce – is constrained by some upper limit. The freedom to introduce some commodities new to the business is also restricted by the need to purchase some form of production entitlement from an existing producer: for example, milk and sheep quotas. 'Set-aside' (the requirement to take some land out of production) is used to limit arable cropping, and the level of income support or compensation received is often conditional on satisfying this requirement.

Product Prices

Product price manipulation is central to most countries' agricultural policies, and many also intervene directly to determine consumer food prices. A major marketing function is thereby effectively transferred away from producers of regulated commodities.

Prices may be modified in the attempt to affect output levels, but more commonly as a means of influencing farm incomes. The level at which the price is set may be manipulated to expand or decrease production of a given commodity, and is affected by costs of production, market demand etc., and by reference to other commodities. Most countries therefore produce a schedule of commodity prices designed to achieve short-term, long-term, or permanent shifts of production from one commodity to another.

The price determined by government, especially where surplus production is the problem, is generally a floor price which provides a safety net for producers faced with a price collapse. The significance of this price to the farmer depends on its relation to the cost of production. If it exceeds the production cost, the farmer cannot lose if he attempts to maximise his output (subject to any existing quantitative restrictions). If the floor price is below the production cost, he needs to decide how much to produce, bearing in mind the perceived risk of a price collapse and any other risk reduction strategies he can employ.

Support prices may also be subject to variation by region (spatial variation), product quality, or time of sale (temporal variation). Regional and quality price variations are generally allowed to express themselves without restraint in developed countries, since they are believed to be useful in guiding production into regions with a comparative advantage in a certain commodity. Quality premiums and penalties are also usually seen as beneficial both to producers and consumers, and are therefore generally encouraged by policy-makers.

Temporal variations, by contrast, are widely used to ensure year-round (or relatively stable) supply of commodities whose production would otherwise be deterred by the high risk inherent in producing and storing them. In practice floor prices are usually varied seasonally, at least to the extent that storage costs are covered. For livestock products the object is to modify production/marketing cycles in an attempt to match supply fairly closely to demand, thereby preventing a build-up of surpluses and associated price collapses, or rapid price increases resulting from short-term deficits. Long-run price stabilisation may also be attempted via the price mechanism.

Production Method

Government action to influence production methods has increased significantly over the past decade in response to public concern about food quality, animal welfare, and the environmental impact of farming. Policy measures may take the form of legal controls and prohibitions on all businesses which are enforceable through the courts, and/or incentives to adopt approved management methods.

All these measures represent a possible marketing constraint through their impact on production: for example, permitted size of poultry cages; a ban on crate production of veal calves; restricted use of veterinary medicines (reflecting animal welfare concerns and human health fears associated with chemical residues in the food product). The storage and application of agrochemicals is also covered by law. Such regulation is, by any standards, a management constraint, and in extreme cases may prevent a business development by entailing unacceptably high financial cost or levels of technical compliance. On the positive side, it has created opportunities to supply specialist markets: for example, organic foods produced without agrochemicals; livestock products produced under guaranteed minimum husbandry and welfare conditions.

> ### Nice food if you can afford to grow it
>
> Hard evidence points to the fact that organic foods are increasing in popularity – and are here to stay. Whether stimulated by food scares, vegetarianism, ethics or desire for purity, the public is putting retailers under increasing pressure to provide them. It sounds like a golden opportunity to cash in ... But UK producers just can't keep up with current demand – and cost considerations are making them reluctant to convert land to organic crops. Major supermarkets are forced to import about 80% of their organic fruit and vegetables due to 'insufficient UK production'. And importing from abroad forces up prices by 25% to 35% in some cases. Only 800 out of Britain's 100,000 farmers are organic, with the total UK market currently worth £200 million. A new study predicts the European retail market for organic vegetables, currently at £122.8 million, will reach £313.3 million by the year 2003 and is likely to prosper as a result of free trade.
>
> *The Grocer, 8 February 1997*
>
> **Box 4.1**

ENVIRONMENT AND RURAL POLICY

The last decade has seen a progressive tightening and extension of regulations governing the impact of agriculture on the natural environment. These policies generally emerged piecemeal, in response to environmental lobbies, public opinion, and the need to restrict the escalating cost of agricultural support while simultaneously maintaining farm incomes. Some measures compensate farmers for practices which reduce their income but are demonstrably beneficial to society. Some directly encourage the adoption of less intensive agricultural methods and conversion to organic farming systems; others pay farmers for agreed countryside and habitat management.

There has also been an extension of rural development policies which have created incentives *and* obstacles to alternative business development. Most governments have policies to encourage rural business development, for instance, but restrictions on land use frequently act to discourage them. Rural businesses are also disproportionately subject to public opinion and organised lobby groups whose intervention influences countryside policies and may block developments.

Integrated policies are now emerging which seek to reconcile these divergent policies, but their implications for the farm business remain problematical. For example, incentive-led environmental management schemes require management time and resources which might be better employed in marketing or business diversification. In many family businesses where labour and capital are limiting resources, this can present difficult management choices and may actively inhibit better marketing.

THE WIDER POLITICAL ENVIRONMENT

In addition to sector-specific policies, farm businesses are subject to political factors which affect all business activity. The political environment is notoriously difficult to predict because it is the expression of public opinion directly through the media and the ballot box, and more continuously through the influence which this exerts on politicians. This was demonstrated by the growth of 'green' politics and militant consumerism, both of which have emerged as powerful organised political lobbies whose impact on government, industry and business is felt worldwide. UK farmers have also experienced direct marketing action by animal welfare and environmental lobbies (who prevented live-sheep exports, for instance, and boycotted products and producers). The same factors, conversely, presented new marketing opportunities: 'green' markets for organic, eco- or animal-friendly products; 'politically correct' markets for Third World products.

In the past, political ideology and party affiliations were considered a reasonably reliable indicator of likely policy shifts with regard to agriculture, land use, and business, but this has become less true as party ideologies have converged in many countries. For most of this century, for example, agricultural marketing boards were an important and stable part of the marketing system in the UK, Australia, Canada and New Zealand, valued equally by parties of the left and the right. This changed in the 1980s, with the emergence of militant free market economics which led to the abolition of marketing boards by a right-wing UK government and by left-wing governments in Australia and New Zealand. Similarly, although the conservative right was once seen as the natural party of agriculture, this cannot now be taken for granted. Party ideology is also not a reliable indicator of how a government or a party will react in day-to-day situations, since this is modified by pragmatic consideration of short-term party interests and domestic and international conditions.

As events in the 1990s amply demonstrated, international politics is one area in which the manager can do very little indeed, although it has the capacity massively to disrupt food markets. International trade is regulated by the World Trade Organisation (WTO, the successor to GATT), which is charged with implementing the 1994 agreement to move towards completely free worldwide access for competing products – including, for the first time, agricultural products. National interests and regional trade groupings will nevertheless raise new obstacles to free trade and genuinely international marketing. Political issues unrelated to trade will also continue to affect export markets and marketing: for example, positive discrimination for products from developing countries; aid for emerging nations; nationalistic attitudes which are reflected in non-tariff barriers to imports (Chapter 13).

THE LEGAL ENVIRONMENT

Although the political environment is notoriously difficult to predict, once a decision arrives in the public arena it can often be modified to meet business interests. The legal environment which is the outcome of the political process is certainly open to modification through the normal consultation processes with which most governments are obliged to comply. Discussion documents about proposed legislation are usually published which invite comments from individuals and trade organisations, and the opportunity which this presents should be maximised because there is little chance of modifying legislation once it is enacted. At the early stage in particular, 'enabling clauses' and provision for local variations can be inserted which allow for unspecified and unforeseeable 'special circumstances' that may subsequently be vital in protecting trade interests.

From a business point of view the relevant areas of the law are those relating to: business form (partnership, plc), taxation and property, premises/facilities and their use, products and their sale. Once a business is established the first two are of little interest to managers, but may be critical when new developments are planned, at which point legal advice is essential. The last two figure prominently in business planning and operation. Farming itself is exempt from planning law in many countries, but non-agricultural farm enterprises are generally subject to it. For example, food processing and retailing enterprises must comply with rigorous construction standards for buildings and plant. An extensive body of law also regulates workers' operating practices, permitted ingredients, product labelling, provision for public access to land and buildings, and so on.

Technical and legal advice on existing legal requirements is widely available, but enforcement agencies are particularly vigilant where the protection of human health and safety is concerned. Since amendments to existing legislation are frequently made under provisions which allow for emergency intervention, producers must be alert to potentially damaging changes which are often introduced without much publicity under existing legislation (for example, changes in packaging regulations; amendments to an EU directive which would have closed all small abattoirs).

Trade associations provide particularly valuable support in monitoring and informing members of such changes, and in representing members where there is doubt in the application of the law. Most of the legal framework may be precise and clear, but there is often room for different interpretations of the same law; it is therefore advisable to take professional advice before embarking on any ground-breaking development.

Finally, it is important to stress that although the law tends to be viewed solely as a business constraint, it also protects the law-abiding against competition from less scrupulous managers. For instance, EU law gives a legally binding definition of free range chicken production, defines standards for organic produce, and protects product labels of origin, to the benefit of the genuine producer. Changes in the law may also facilitate business developments.

Packaging regulations loom

Two out of three food and drink firms are unaware of the new packaging laws before Parliament this week. The Producer Responsibility Obligations (Packaging Waste) Regulation sets out measures to reduce eight million tonnes of packaging waste each year. The Government's environment committee will agree or reject the legislation within four to six weeks.

The Grocer, 8 February 1997

Box 4.2

THE ECONOMIC ENVIRONMENT

The economic environment can and does change very rapidly and often requires changes in marketing strategy. A perfectly sound marketing strategy which would succeed in a period of average growth, for instance, may fail in one of economic stagnation or recession.

The *macro*-economic situation affects businesses and customers, and has international repercussions. Customer attitudes towards goods and expenditure are affected both by their view of the current economic situation and their expectations of future economic performance (with particular concern for employment prospects). Both consumers and producers are affected by direct and indirect tax levels and interest rates, which affect producer, manufacturer and retailer costs, investment levels, production targets and sales. Governments and central banks use economic regulators like tax changes, interest rate changes and intervention in foreign exchange markets in the attempt to steer the economy in a desired direction, and managers can only try to anticipate the likely effects. In the longer term it is possible to seek to influence the process through trade and professional organisations and elected politicians.

The *micro*-economic situation affects long-run marketing strategies and short-run marketing activities through its impact on the consumption and production of goods and services, their costs and prices, and the competitive framework.

Consumption is obviously affected by micro-economic factors, of which price is always a major preoccupation which periodically overrides all others. Governments may be very sensitive to consumer complaints about exploitation or over-charging, and may intervene to make firms reduce prices, or threaten an investigation which may achieve the same objective. In a situation of monopolistic or oligopolistic competition, businesses may choose not to compete on price, but on terms of sale (money back with second purchase, promotional give-aways, product modification etc.). This is seen as being in the consumer's interest, and as a driving force in business efficiency and innovation; activities which are misleading (if not actually fraudulent) may nevertheless provoke government intervention to protect the consumer by modifying the terms of competition.

The *competitive structure* of agriculture is strongly influenced by the fact that it is an atomistic industry, composed of large numbers of producers competing with each other in the production of largely homogeneous goods. In this situation every farmer is in theory a price *taker*, who may achieve no sales if he seeks a higher price for an undifferentiated product. In the food manufacturing and distribution sector, by contrast, a few very large firms are often price *makers*, and a price increase does not result in total loss of trade (Chapter 8). In practice, businesses can adopt a pricing policy to put pressure on competitors: a business with substantial resources may cut prices below cost to squeeze out competitors, then raise prices without fear of losing sales. It may also buy out other businesses to the same end. Farmers can form joint ventures which have the same effect of reducing competition.

All these considerations are complicated by international politics, which is particularly relevant to exporters, to businesses dependent on imported components and raw materials, and to domestic producers of raw materials faced with competition from abroad. The two critical issues to be considered are trading access and exchange rates, which have strong political as well as economic connotations.

Trading access

Economic theory can make a case both for free trade *and* for the protection of a country's industries by import restrictions. The free trade argument is currently dominant, as formalised in the 1994 GATT agreement which requires increased access for agricultural imports. In the EU this will entail a reduction of production levels (since consumption is not expanding) and a reduction in subsidised exports from the Community to third countries. A review of the agreement in 1999 is expected to lead to further liberalisation and dismantling of formal barriers to world trade, but some observers see signs of a weakening commitment to free trade on the grounds that Third World interests are damaged by it, as large exporting countries use the agreement to their own advantage.

The implementation of global free trade is also obstructed by regional trading areas which have free internal trade inside a tariff wall to keep out competing products: for example, the EU, NAFTA (North American Free Trade Area), Mercasur (South American countries). Even within supposedly free trading areas a wide range of non-tariff barriers exist which inhibit trade: for example, non-common product specifications; different phytosanitary standards. The EU Single Market has in theory removed most barriers of this nature, but they still exist, and exporters may have to make rapid and imaginative responses to cope with new ones (Chapter 13).

US seeks an unfair advantage

The new US agriculture policy has serious consequences for the EU's food and farming sector. It is likely to lead to greater restrictions on export subsidies ... and force the EU to change the way it supports markets. At the same time it will give the US greater freedom to aid its food producers without breaching WTO rules. While maintaining subsidies to producers it will allow output and exports to expand without limitation ... The new legislation is presented in the US as a major liberalisation of food and farm trade which will stabilise world markets ... Instead the US is seeking to establish both an unfair advantage in world markets and to encourage if not force other countries to change their own policies towards greater liberalisation. This may unleash some form of retaliation, leading to greater volatility of markets.'

The Grocer, 30 November 1996

Box 4.3

Currency exchange rates

Currency exchange rates are a major management preoccupation for export marketers, though there is little they can do about them. Exchange rates may be floating or fixed. (EU farmers are also familiar with 'green currencies', but these play no part in commodity trade, being used only to convert support prices into national currencies.)

Countries with *fixed exchange rates* decide as a matter of policy what the conversion rate of their currency should be in relation to one of the major trading currencies (US dollar, D-mark). All trade is then undertaken at this rate. If the currency is convertible (can be exchanged by anyone for any other currency) its 'value' will be determined by its supply and demand, and trading in the market allows the government to defend its chosen exchange rate through open market operations. If a currency is felt to be overvalued the market allows selling, which lowers its value and may force government to devalue to a new lower level.

Fixed rates exist for non-traded (non-convertible) currencies in Eastern Europe, Russia, and most developing countries. These cause problems for traders wishing to sell into these markets, so potential exporters must take specialist advice. Competitor businesses in countries with fixed exchange rates may also constitute 'unfair' competition, since the country may charge any price it wishes in order to sell products as a means of gaining foreign currency earnings (a common policy in the past which should diminish as trade liberalisation develops).

Floating exchange rates were unheard of before the 1980s, but they are now the norm in developed countries. In this situation currencies are allowed to float in line with the market, which makes trading difficult for small businesses, who have to allow for exchange rate fluctuations in costing, pricing, and viability assessment.

Within trading blocs there is a move towards a single currency (the EU Euro, for example) to eliminate this problem in internal trading. In Canada, Mexico and the USA the dollar is effectively such a currency, which benefits international traders by reducing transactional costs and the need for protection against exchange rate risks. On the negative side, a substantial one-off cost is incurred in converting all the purchasing and selling systems throughout a single currency area.

THE SOCIO-CULTURAL ENVIRONMENT

The socio-cultural environment affects the market for a product, business structures and practices, and the political, economic and legal structures in which firms operate. In most societies the socio-cultural environment is in a state of constant change to which businesses must adapt, though businesses can (and often do) set out to change social habits, lifestyles, and even cultural beliefs, to their commercial advantage.

Two categories of socio-cultural factor may be distinguished:

- demographic factors which are reflected in relative spending priorities, buying patterns, urban/rural variations etc. (population size, structure and distribution, including age structure and regional distribution)
- less tangible factors like community values, beliefs, and attitudes, class structure and customs, educational level, the status and role of women, religious influences, ethical values etc., which affect how and why people live and behave as they do.

Demographic data are fairly easy to obtain, and their significance for the marketing mix is relatively easy to identify: for example, the relative distribution and spending power of the single elderly, or of young single households. The second group of factors are more difficult to identify and interpret, but the attempt must be made because they frequently determine the viability of a marketing programme. It is also important to understand the ways in which cultural and social customs are transmitted through society: by personal contact, by education and, increasingly, via the media and information technology, since this helps in designing the promotional aspects of a marketing programme, and may be critical in gaining market access.

The implications of socio-cultural factors for the marketing mix are discussed further in Chapter 6. Their possible impact is simply indicated here by reference to one factor – the educational level of customers, which includes receptiveness to information as well as formal education. For instance, it may be obvious that a product whose successful use requires careful reading of instructions is unlikely to sell well to consumers with a low level of literacy, which may exclude it from some export markets. However, even highly literate consumers commonly disregard manufacturers' instructions, so a product whose use requires detailed explanation faces an information and acceptability barrier in any market.

Product modifications and changes to the marketing mix may therefore be necessary to enhance product acceptability and reduce the potential for misuse which will result in customer dissatisfaction. For example, packaging design will have to provide clear on-pack instructions and still remain attractive enough to catch the customer's attention, as Plates 4 and 5 demonstrate. As Plate 5 suggests, however, what seems only a constraint may also be an opportunity. In this case, value is added to a basic commodity through a product assurance and an inexpensive product addition which allows a customer without cooking skills to achieve a better result. The price for the product also rises sharply compared with that of the standard product in the same supermarket cabinet.

One social force which has exerted a particularly strong influence on marketing in the last decade is the impact of organised political lobbying by consumer groups, environmental lobbies, ethnic and religious interests etc. These groups exercise a strong influence on consumer opinion, and have stimulated extensive government intervention in all business, not just agriculture. The result is a substantial body of legislation which now protects consumer health, welfare and legal rights, with which managers must simply comply. Equally important to managers who want to stay in business is the moral pressure brought to bear on businesses to act 'ethically', which has been especially critical in the development of less intensive agricultural regimes. Once seen as a constraint, this is now seen as an opportunity for 'ethical marketing' which stresses environmental responsibility, animal welfare, healthy living etc.

Portraying an ethical image

More than two thirds of adults now consider a supermarket's ethical and social stance when buying products. Some 86% agree they have a more positive image of a company if they see it doing something to make the world a better place, and 60% of shoppers are willing to boycott products or stores on ethical grounds.

The Grocer, 14 January 1997

Box 4.4

THE TECHNOLOGICAL ENVIRONMENT

The impact of technology has extensive implications for the product and marketing mix, and may be a major factor affecting product/business viability and profitability. Processing equipment for a farm-based food enterprise may rapidly become outdated by a technological innovation with which small processors cannot compete, or by modified legislative requirements which reflect the availability of new technology they cannot afford to install. Farmhouse self-catering accommodation without the latest electrical appliances may be difficult to market, although the financial implications of equipping and regularly re-equipping units may not be justified by projected revenue.

The speed of technological change in the last two decades has accelerated dramatically, and tends to shorten the profitable life cycle of products, entailing more frequent product modifications and innovations in order to keep pace with customer expectations and competitor offers (Chapter 7). In the food industry especially, production-line technology and quality control, coupled with information technology, have dramatically improved the manufacturer's ability to make and insist on very precise input specifications to achieve maximum technical and financial performance. The meat processing sector, for instance, was radically transformed by the advent of primal cutting of carcasses and controlled atmosphere packaging, which had radical implications for product specifications and livestock production systems.

Equivalent advances have been made in marketing logistics at all levels, with similar effects on suppliers. Faster and cheaper transportation in the early 1990s allowed long-distance air-freighted supplies to compete against domestic supply sources. This increased both the quantity and variety of available products, and the level of competition in the market: for example, chilled (not frozen) New Zealand lamb which competes directly on product quality and price with fresh UK supplies.

At retailer level electronic technology has provided near-instantaneous stock control and sales monitoring, greatly enhancing the retailer's ability rapidly to de-list products which do not sell. At consumer level, the introduction and increased ownership of home freezers and the microwave permitted a massive expansion in innovative processed food products. More recently, wider ownership and use of home computers has increased the potential for direct marketing, and the substitution of home shopping for the personal store visit.

The implications of technological advances at every level of the marketing system must therefore be monitored, as well as technological advances in production. Managers must also be aware of the generally rising technological expectations of their customers, although there is also a market for low technology which farmhouse entrepreneurs frequently exploit: for example, 'natural' food in unsophisticated packaging; cheap bunkhouse accommodation; 'traditional', low-tech leisure activities. A careful balance of high and low technologies is invariably needed, however. Customers for traditional farm accommodation expect efficient central heating and facilities comparable to those they have at home. Buyers of 'natural' food expect (and there is a legal obligation to provide) a product made in accordance with the latest health and hygiene standards, frequently using sophisticated technology.

THE MICRO ENVIRONMENT

The micro environment is conventionally defined as being a firm's customers, distributors, competitors and suppliers. The remainder of this chapter deals briefly with competitors and suppliers; distributors and customers receive more detailed consideration in subsequent chapters.

COMPETITORS

Although every firm would like its customers to believe that it provides a unique product or service, in practice every business and every product has a competitor. Even where a product seems to be unique, substitutes invariably exist – nowhere more so than in the food sector, where the choice available to consumers in developed countries is limitless.

Managers can exercise some control in this competitive environment by adopting strategies which avoid head-on competition, and they can plan to meet competition where it is unavoidable by identifying new or better ways of supplying a market. To do this they need information about:

- existing and potential competitor products and their market positions (distinct product characteristics; customer segments targeted; neglected or weak segments)
- competitor strategy and behaviour (for example, market share growth via price competition, or long-run profitability via a quality product and added service)
- their competitive and management strengths and weaknesses (for example, non-innovative; strong resource base; weakness in marketing or production).

It is important to distinguish between *inter-firm* competition (Ford v. Nissan, Farmer Jones v. Farmer Brown) and *inter-product* competition (meat v. fish/pasta/ready meals; fresh v. frozen vegetables; 'natural' v. processed foods).

Inter-firm competition

Producers of mainstream farming commodities have been shielded against inter-firm competition by subsidised prices, and it is still arguable that the farmer's chief competitor – in terms of total consumer expenditure and his share of it – is not other farmers, but the rest of the food industry. As agricultural markets are liberalised, however, inter-farmer competition is increasing rapidly within national boundaries as well as at the international level. Fine judgement is therefore required in deciding where and when individual competitiveness is better served by more aggressive individual marketing, or by joint ventures with other producers.

The problem lies in the atomistic nature of agricultural production, which is characterised by many small producers of relatively homogeneous products such as livestock or wheat. This makes it almost impossible for even the largest individual producer to supply the needs of markets which are characterised by a few very large buyers (Chapter 5). The individual's bargaining strength is also negligible in markets which are often grossly oversupplied. Cooperation with other producers (*horizontal coordination*) may therefore be the only way to supply sufficient volume to meet a specific demand, and a more appropriate competitive strategy than stronger inter-farmer competition.

In non-farming enterprises, inter-firm competition quickly becomes apparent as existing operators move to resist an incomer, so it is vital at an early stage to identify exactly who is the competition and exactly what is their offer. Identifying *existing* competitors is relatively simple, and it should be possible to identify the strengths and weaknesses of their offer by observation, by sampling their product, and by questioning customers. It is much more difficult to calculate how competitors will react to new competition: for example, will they respond with aggressive price discounting and promotional offers? In the case of large competitors it may be possible to deduce this from their reputation and public statements about their management objectives and strategy (annual reports, newspaper accounts, promotional material etc.). For smaller competitors, including other farmers, it may have to be deduced from past behaviour and present management style.

The identification of *potential* competitors is an even tougher exercise which many entrepreneurs overlook in their enthusiasm for a new product idea. Few ideas are really unique, however, and many can be copied fairly easily. Indeed, the one thing small innovators may be sure of is that their innovative idea will be copied at the earliest opportunity, and will probably be bettered by established firms who have the resources and determination to defend their market position.

Inter-product competition

Inter-product competition is especially important in the agrifood sector, where the basic product (an 'eating experience') is available in a variety of forms limited only by the manufacturer's ingenuity (Plate 6). At the most direct level competition exists between identical or near-identical products: feedwheat, baked beans, bread. In this situation products compete in exactly the same market, and differences between different brands, labels, or varieties are limited to a small range of characteristics which are typically not intrinsic to the product. Competition is therefore on price, added benefits, and brand identity.

Peas offering

Quorn and tofu have a new rival. A pea and wheat protein substitute under the name Arrum has started to appear in ready meals ... Birds Eye Walls has been the first to take it on board, incorporating it into one of its four meat-free ranges. But we will not be seeing Arrum flagged under its own name.

It will be supplied as an ingredient to branded and own-label manufacturers for use in frozen, chilled and canned products. Ready meals, pies and pasties are to be the prime target. Arrum has a fat-to-calorie ratio of 11% compared with 54% for beef and 32% for chicken ... Birds Eye Walls' first uptake is as a beef substitute.

The Grocer, 17 August 1996

Box 4.5

This form of competition is the commonest in most businesses, and certainly for farmers who produce undifferentiated commodities like milk, main crop potatoes or cereals, where any differences relate only to crop varieties or livestock breeds and their characteristics and the farmer's reputation as a supplier (brand). It is seen most acutely in the supply of agricultural raw materials for processing, where substitutes – including other agricultural raw materials – increasingly compete on price and technical performance.

Finally, it is important to remember that competition from all other products as well as food must also be considered, since they compete for consumer expenditure. Consumer expenditure priorities vary widely from one country to another, but well-fed consumers frequently economise on food rather than on fixed expenditure like housing or discretionary spending on leisure, entertainment, and holidays. This is one reason why added-value food products may not necessarily bring the additional revenue expected, since some consumers cannot afford or do not see expenditure on added-value food as justifiable. Investment in a non-food enterprise (say an accommodation or leisure facility) may therefore yield a higher return than an equivalent investment in a food-production enterprise.

SUPPLIERS

The reliable, timely and cost-effective procurement of production inputs is critical to efficient marketing. The inability or unwillingness of suppliers to provide inputs as required can disrupt production and marketing schedules and result in lost sales, lost customers and unachieved planning targets. The quality of inputs is also crucial to a quality marketing proposition, as the BSE crisis demonstrated, when animal feed was identified as the likely source of the disease. In the past, producers tended to take input quality for granted; today market access is increasingly recognised as dependent on the ability to prove effective quality control over production inputs, as a guarantee of output quality.

Access to supplies

For small businesses diversifying into new enterprises it may prove difficult to identify sources of supply in unfamiliar business territory, so the temptation exists to seize on the first supplier. It is nevertheless vital to ask all the right questions at the planning stage, because failure to procure the right inputs may invalidate an otherwise sound project. For example, how do you locate reliable supplies of organic fruit for organic yogurt, or yogurt pot manufacturers and designers? Do they have minimum order sizes? Can they guarantee to supply a design over the next three years, and expand supply to meet an expansion in demand? What alternative is there if they go out of business overnight? Are the pots recyclable?

These problems can be solved where many potential suppliers exist, but shopping around costs management time which an owner-manager cannot afford. Where there is only one supplier (actual or known) insecurity of supply is a serious risk, and an absolute refusal of supplies is not unknown. A change of production may therefore be unavoidable, which may simply mean modifying a production method or accepting a lower quality or yield. At worst, the abandonment of an enterprise may be unavoidable: for instance, for some crops access to irrigation is indispensable to achieve quality and quantity output targets.

Input price rise effects

Any supply shortage may result in rationing and will usually result in a price rise, and the combined effect of rationing and price increases will reduce profits. A price rise inflates costs but may leave revenue unaffected. Non-price-related rationing is likely to reduce revenue, but may be accompanied by a slight reduction in costs. The decisions which the manager needs to make in these situations may be quite different, although both will involve some purchasing, production, and sales and marketing changes. The range of options available will depend on the input concerned, the nature of the production process and the market situation. It will also depend on an assessment of the likely duration of the shortage, since a temporary shortage will require less radical business reappraisal and restructuring than a permanent one.

Faced with an input price rise, the questions which need to be answered are whether:

- the input is indispensable to the production process
- the amount used can be reduced (by reducing wastage, for instance)
- a cheaper substitute can be found which will maintain or reduce costs without adversely affecting the quality of the product or the production process.

In a well-managed business none of these options ought to be available, since a well-managed business should not have to wait for a price rise to identify improvement potential. The market situation will determine whether the price rise can be passed on to customers or must be absorbed by reducing the profit margin. The ability to raise the customer price depends on a range of factors, but chiefly the nature of demand (budget-conscious or relatively affluent), and competition in the market, both from other producers and from substitute products.

If an input price rise is likely to be sustained in the long term, a series of more strategic issues arises, involving reappraisal of the production process. Can the process be modified or the product reformulated so that it uses a smaller quantity of the input, or a substitute input be used without affecting demand? In an agricultural enterprise, can yields be reduced while maintaining profitability? Can the total product mix be restructured to ensure that the more expensive input is used for the most profitable output?

A similar reaction is likely in the face of administrative controls, but in this case the strategic options will need earlier consideration, and every possible means of ensuring the most effective and profitable use of the scarce resource must be considered. If the rationing is likely to be permanent, the search for alternative reliable supplies may include the possible acquisition of a supply directly under the business's control, and the search for substitute inputs will also be more active.

The nature of the input is relevant to these decisions. If it is a common packaging material or a particular brand of fertiliser or agrochemical, an alternative, lower-cost supply source may be identifiable. If there is no substitute and the input is critical to production (for instance, water for crop irrigation, fish farming, or farmhouse processing), the increased price must be compared with the benefits derived from its use. Given these two figures, it is relatively easy to decide the right strategy: to make do without the input and reduce costs and revenue (lower saleable quantity and quality), or to absorb the cost from profits or pass it on to the customer. If the input is indispensable to product quality, the latter may be the only realistic solution once production is on stream and the producer has a reputation and a market which must be defended. However, promotional activity will be necessary to defend the price rise and prevent lost sales.

CONCLUSION

The marketing environment is the major source of planning uncertainty for any business, but good marketing intelligence can reduce it to manageable proportions. At the very least this means identifying existing and likely constraints which must be taken into account in planning the marketing programme. More positively, it requires an alertness to external conditions which will mitigate their worst effects, and may allow a business to capitalise on unexpected opportunities.

CHAPTER 5

Understanding the marketing system

Products are distributed to the consumer via marketing channels which may be more or less direct, formed by the trading activities of firsthand intermediaries, processors, wholesalers and retailers, who buy and process raw materials and distribute finished products to consumers (Figure 5.1). In addition to these core marketing institutions, a wide range of commercial firms provide essential services to the system although they are not part of it: banks and insurance companies who provide capital and risk-sharing facilities, transport companies, advertising and PR agencies etc.

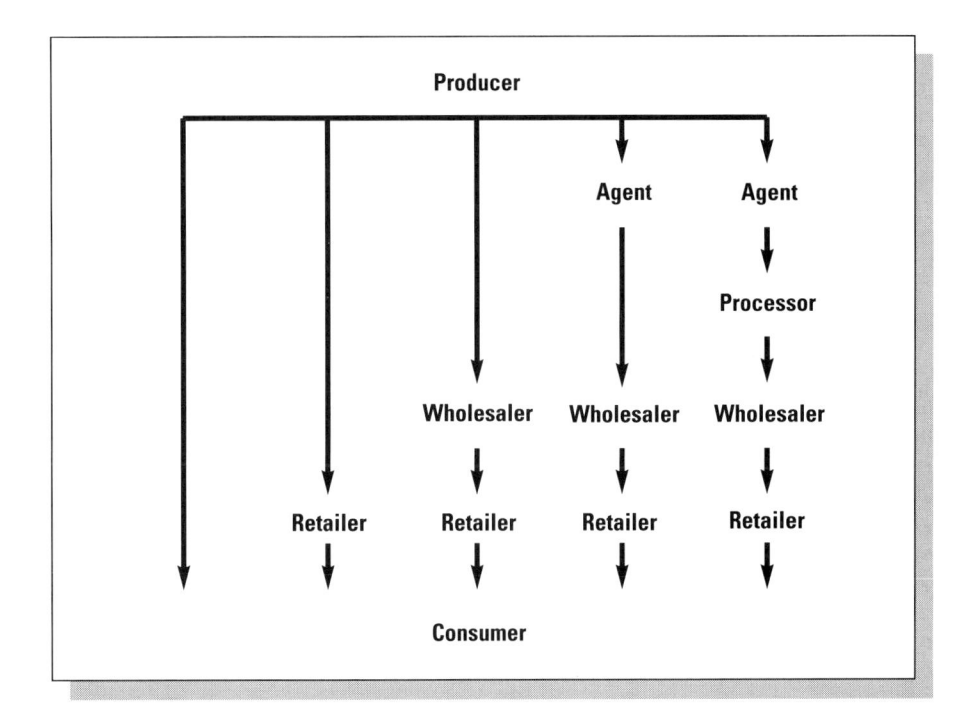

Figure 5.1 Marketing channels

The resulting product marketing systems are very variable over time and place, depending on environmental circumstances (geography, socio-cultural and legal influences and the historical development of the channels) as well as product characteristics and industry sector. All marketing systems, however, perform the same broad functions:

- *logistics*: everything involved in moving a physical product (transportation; warehousing; sorting and grading; storage etc.)
- *processing*: product transformation, presentation and packaging

- *communications*: product ordering, customer research and feedback, promotion
- *economic functions*: primarily the rate of exchange (price), but including financial services (credit, insurance etc.).

The simplest marketing system is the producer-retailer who single-handedly produces, processes, and distributes a product to the end-user without intermediaries: for instance, a farm shop, doorstep milk delivery, craftsmen selling direct to the public, mail order. At the other extreme is the agrifood marketing system of a country or the world, which performs identical functions, but via exceedingly complex channels – reflecting the fact that it must supply millions of consumers daily with a huge range of fresh and processed products sourced from almost every country in the world (Figure 5.2). In many countries it is also the biggest business sector and the largest employer in the economy, and a major export earner. (The value of products flowing through the system in the UK is shown as an example in Figure 5.3, which (though simplified) indicates the size of the major sectors and employment generated.)

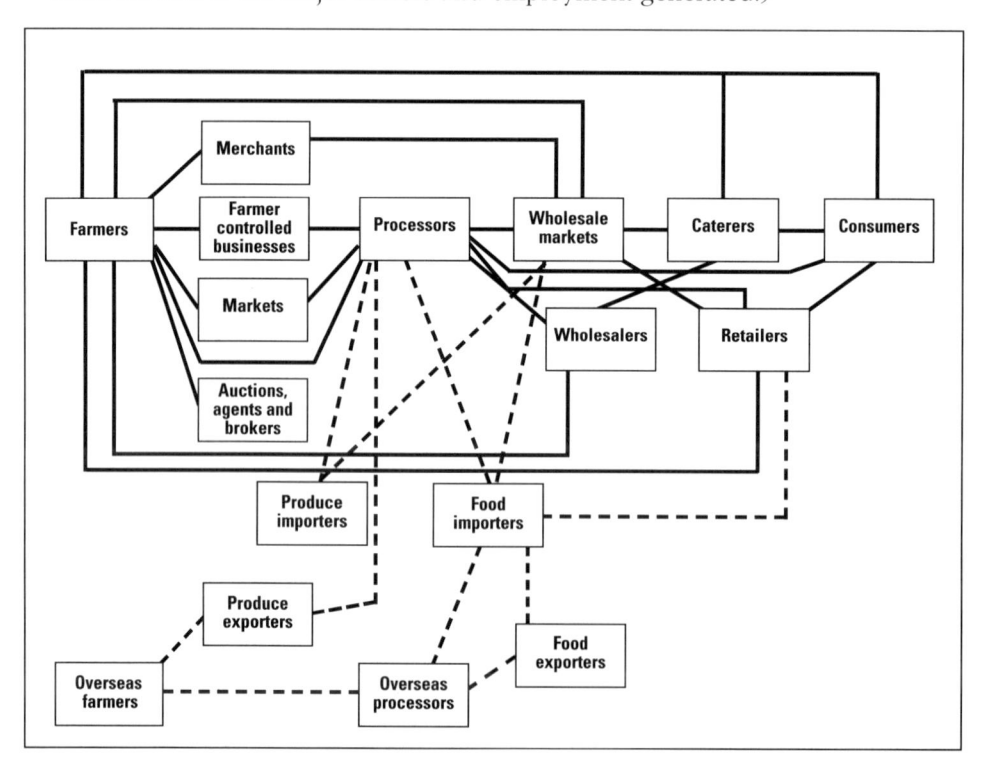

Figure 5.2 The agrifood marketing system

This chapter concentrates on the agrifood marketing system because that will be the reader's main interest, but the implications for other product sectors may be readily inferred because the account focuses on functions rather than institutions, and other systems are invariably much simpler. In most agrifood marketing systems three kinds of marketing institutions are recognisable:

- *firsthand buyers* who organise the sale of primary commodities to buyers further downstream (merchants, auction marts, producer marketing groups etc.)
- *processors* who transform raw materials into finished or part-processed products (flour millers; oilseed crushers; abattoirs; manufacturers of meat, bakery and frozen food products)
- *distributors* who make products available to consumers (wholesalers, retailers, caterers).

These divisions are not water-tight, for many firms and functions over-lap. The small milk producer-retailer, the retail butcher who slaughters and sells his own meat, and the huge multiple retailer chains (multiples) are all involved in processing and distribution. The dividing lines have also become blurred in many countries as both ends of the marketing channel seek to cut out the intermediary functions of the 'middleman'.

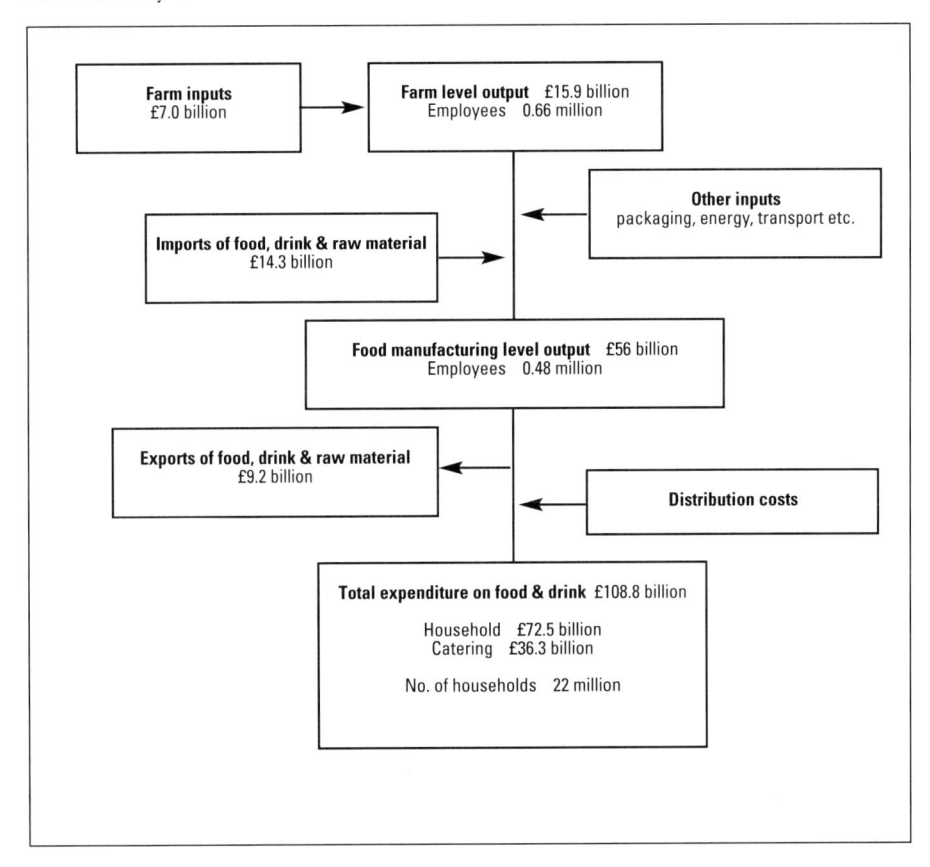

```
┌─────────────────┐      ┌──────────────────────────────────┐
│ Farm inputs     │─────▶│ Farm level output  £15.9 billion │
│ £7.0 billion    │      │ Employees          0.66 million  │
└─────────────────┘      └──────────────────────────────────┘
                                            │
                                            ▼
                                 ┌──────────────────────────────┐
                                 │ Other inputs                 │
                                 │ packaging, energy, transport │
┌──────────────────────────────┐ └──────────────────────────────┘
│ Imports of food, drink & raw │
│ material  £14.3 billion      │
└──────────────────────────────┘
                │
                ▼
      ┌──────────────────────────────────────────┐
      │ Food manufacturing level output £56 billion │
      │ Employees  0.48 million                  │
      └──────────────────────────────────────────┘
                │
┌──────────────────────────────┐      ┌──────────────────────┐
│ Exports of food, drink & raw │◀─────│ Distribution costs   │
│ material  £9.2 billion       │      └──────────────────────┘
└──────────────────────────────┘
                │
                ▼
      ┌──────────────────────────────────────────┐
      │ Total expenditure on food & drink £108.8 billion │
      │                                          │
      │ Household  £72.5 billion                 │
      │ Catering   £36.3 billion                 │
      │                                          │
      │ No. of households  22 million            │
      └──────────────────────────────────────────┘
```

Figure 5.3 The value of products flowing through the UK agrifood system 1994 (latest data)

For a marketing system to work effectively and achieve the objectives of everyone involved in it, all parties need to work towards the same goals: there must be effective *channel coordination*. The ultimate shared objective is to satisfy the consumer who keeps everyone in business, but every business has its own objectives, which results in competition to supply its part of the marketing task. What the producer needs to find is a route through the system which delivers the best collective benefit as well as satisfying his own objectives.

The price which a marketing channel currently returns to the business should rarely be the only objective: it is more relevant that the buyer is selling into markets which will continue to pay a good price because they accurately express end-use demand, and are innovative. This point was brought home to many farmers for the first time in the 1990s by the abolition of statutory marketing boards which had previously taken responsibility for the disposal of many mainstream agricultural commodities. UK dairy farmers, for example, suddenly found themselves obliged to identify and evaluate alternative milk buyers without enough marketing intelligence to make an informed choice.

Producers realised that the crucial consideration was not a firm's opening price, but the stability and growth potential of the markets in which it operated, and its efficiency and innovative record in supplying changing consumer demand, since these were a better indication of the price it could *continue* to pay for milk. However, few farmers knew enough about consumer markets or the downstream sector to evaluate the merits of a fresh products company producing retailer own-labels, a firm with its own brand, a doorstep delivery firm, or a multi-product company, nor which was likely to be a better long-term marketing partner.

It is clear, then, that even the apparently simple choice of a firsthand buyer requires a knowledge of the marketing system, because end use and the nature of the marketing channel will affect the revenue earned and the quantity and quality of marketing intelligence received – and hence, future revenue. It may also affect the amount sold and long-run marketing prospects (if, for instance, a buyer/channel loses touch with end-use demand or the confidence of buyers further downstream).

The management considerations involved in the distribution decision are discussed in Chapters 9 and 10. The present chapter outlines the strategic and commercial motivations which drive the processing and distribution sector, and thereby determine the product and marketing specifications suppliers have to meet. The account focuses on the UK, but the system is representative of the agrifood marketing system in most developed countries, which is rapidly becoming the global norm.

PROCESSING

Processing adds form and utility to raw materials by adding services which deliver a finished product consistent with end-use requirements. In most countries the food processing sector (the 'food industry') is intensely competitive, and business survival depends critically on two factors:

- the capacity to respond to changing consumer demand ahead of the competition
- the ability to purchase raw materials from primary producers and sell to retailers at prices which maintain manufacturing margins.

The demand pressure is intensified because it comes not from the consumer, but via a retail sector which in most countries is dominated by a few very large multiple retailers who are locked in bitter competition for sales volume and margins. The price of failure to supply these retailers is too high for the food industry to risk – even though it is itself composed of some of the largest companies in the world. The food industry's only loyalty to suppliers therefore depends on their ability consistently to deliver 'to spec', and technological advances have eroded the advantage which farmers once enjoyed in producing a perishable product which had to be sourced locally.

Raw materials today can be purchased wherever price, quality and availability are most advantageous, and manufacturing operations can be shifted to an optimal cost-reducing site. In the EU single market, food processors routinely source raw materials on a Europe-wide basis, and companies with manufacturing capacity in several countries have moved some processing operations to achieve their price and cost objectives. The reverse of the coin is that all food processors in a single market are a farmer's potential customers, and marketing across national boundaries has become an increasingly common option.

This has particular potential for UK producers, since the domestic food market is comparatively static, reflecting a slow rate of population growth and the fact that most people are more than well fed. Penetration of the home market by foreign suppliers has also made deep inroads into UK consumption. Simply to stand still the UK industry must therefore try to raise revenue by continual innovation and cost cutting. This shows itself in increasingly high specifications for raw materials, in the drive to reduce wastage and raise food quality (which often means less highly processed foods, requiring even higher input quality).

INDUSTRY STRUCTURE

The structure of the UK food industry is shown in Figure 5.4, with firms arranged by number of employees. The vast majority of firms are quite small (90 per cent of firms have less than 99 employees, and 60 per cent have less than nine), but these firms together contribute only 14 per cent of sales. The

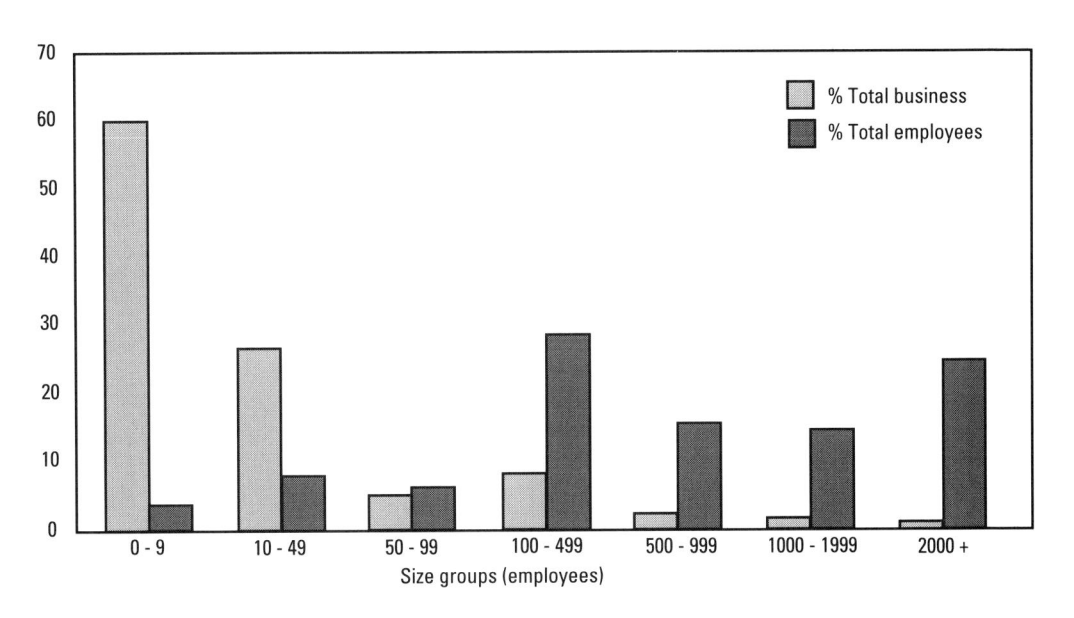

Figure 5.4 Structure of UK food industry by employment generated, 1994

largest five companies, by contrast, with more than 2,000 employees each, contribute 24 per cent of sales output.

This picture of a concentrated industry, dominated by a handful of large companies, is reinforced at the sectoral level (Figure 5.5), where the largest five firms generate more than 50 per cent of total sales. Some of these companies (usually large multinationals) have a very wide product base, while others are very large in a narrow area in which they specialise (sugar, dairy, meat). In between there is a large group with a fairly wide product range (dairy, meat, bakery products, frozen foods etc.).

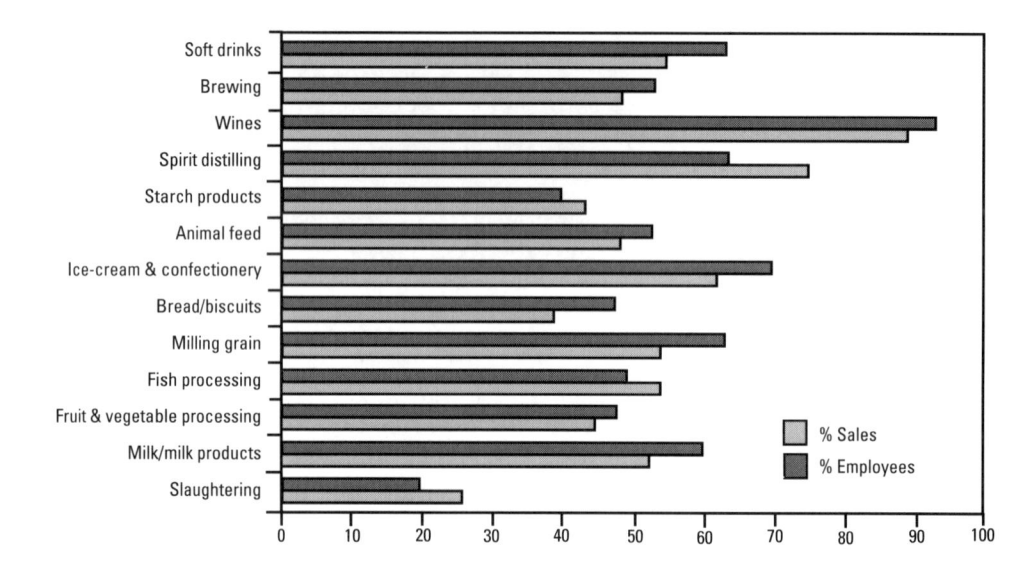

Figure 5.5 Share of industry employment & sales accounted for by largest 5 firms

The progressive concentration of the industry in fewer hands is a continuous process, and in some sectors has been very rapid: for example, meat. In other sectors the number of buyers has already been reduced to one or two firms, and further contraction would mean that the sector would be entirely lost to the national economy. The trend is nevertheless encouraged in a single market like the EU, which enables a company to centralise all its manufacture of a product or a product group in a single factory from which it supplies the entire European market. It has also been reinforced by advances in technology which permit even larger economies of scale, and by past changes in the CAP and continuing uncertainty about the future direction of agricultural policy. The imposition of national milk quotas, for instance, obliged some firms to manufacture products in several countries in order to obtain sufficient milk. Alternatively, a quota-restricted supply of raw material may lead a company to concentrate its manufacturing operation on the highest value product in each particular market, and tranship its products throughout the EU, allowing it to offer a range of products to any given retailer while simultaneously benefiting from an assured supply.

The small processor sector

At the other end of the scale, the food industry has a large number of small, generally single- or narrow-product range manufacturers, because entry to food processing is encouraged by the relatively low set-up cost. Farm production of territorial cheeses, ice-cream, cakes and specialist meat products is now very common, and some farmer-processors have moved into factory premises to meet expansion targets (Plates 7a and 7b). Farm processors have often led innovation in the industry, and many innovative products and businesses have been taken over by larger firms, leading to concentration in this sector too. Many small processors also fail and leave the industry, to be replaced by newcomers with new ideas or skills.

These innovators face formidable obstacles. For example, small processors have suffered a relative disadvantage in production technology which has only recently been reduced by the advent of new technologies that do not require high production levels to make them financially viable. Selling products to retailers who are bombarded with competing offers becomes increasingly difficult, however, following the adoption of electronic point of sale stock control and product evaluation (EPOS) which allow rapid de-listing of products that do not meet retail profitability targets (Plate 8).

Physical distribution is also difficult for small processors, because they may have to deliver direct to individual retailers (a high-cost and time-consuming undertaking) or via wholesalers (often unreliable). In theory small producers should be able to supply large multiple retailers who have their own central distribution systems, which allow deliveries to a central depot from which it is the retailer's responsibility to distribute products in appropriate conditions to individual stores (particularly important for perishable or fragile products). In practice, access is limited by the exacting demands of the centralised system and the retailer's requirement for high-volume, consistently available supplies. The multiples also frequently want small processors to produce own-label products, sold under the retailer's rather than the producer's name. This benefits the small processor by reducing the cost of establishing and promoting his own brand, but if the retailer's custom is lost, the producer will be left with no customers loyal to his own brand.

FUTURE TRENDS IN PROCESSING

Food processors pursuing a long-run added-value strategy in a highly competitive market need a large, consistent supply of a wide range of competitively-priced raw materials. Since the 1960s there has been no shortage of most agricultural raw materials, but support policies kept raw material prices artificially high to maintain farm incomes, and import tariffs and restraints limited manufacturers' access to cheaper third country supplies.

As a result of the 1994 world trade agreement, competition from imported raw materials will inevitably grow. Competition must also be expected from raw material substitutes which are increasingly more price competitive as well as technically more satisfactory, and now acceptable to a growing number of food consumers. A resulting downward pressure on raw material prices is therefore likely, as intensifying competition on several fronts toughens the resolve of the manufacturers and distributors to maintain their respective margins.

Since the 1970s food manufacturers have complained that their profitability has been sharply reduced by the dominant position of the large multiple retailers, but calls for monopoly investigations were ignored by government because the consumer interest was not seen to have been damaged. In the UK, however, profits were so low in some sectors that many processing firms left the industry (notably, in the abattoir and bread-making sectors), resulting in a concentration within the sector which may disadvantage farmer suppliers. Manufacturing margins recently showed some recovery, but retailer margins declined as competition at retail level intensified, which means that they cannot raise consumer prices. The pressure on manufacturers' margins therefore continues, and will further squeeze the producer's margin as increased access for imports affects world markets.

DISTRIBUTION

Distribution involves a wide range of activities which transfer products from point of production to point of consumption (Table 5.1): assembly, sorting, stockholding, physical transportation and selling, together with all the necessary communications functions which facilitate trading (product ordering, invoicing, credit provision, customer feedback, etc.). Distribution channels and distributors are consequently very product- and sector-specific, and vary widely from one country (even one region) to another. The distribution channels for agricultural and food products are characterised by great complexity which is much simplified in the following account, reflecting the reality on the ground, which is to simplify the distribution function by cutting out unnecessary middle layers.

Marketing theory and many product distribution channels recognise two main organisational layers within the distribution sector:

- *wholesalers*, who assemble products and make them available to
- *retailers*, who make them available to consumers.

This distinction is relevant to the producer of consumer (finished) products, who must understand the relative merits and disadvantages of using wholesalers and retailers, and the implications of seeking to by-pass both and distribute products (as many producers hope to do) direct to the end user (Chapter 10).

At the agricultural commodity level, producers are more concerned with many kinds of firsthand buyers who perform the same kind of assembly and communications functions as wholesalers in other product sectors (Chapter 9). Conventional wholesalers and wholesale markets as such have been in long-run decline in the agrifood sector in many countries, and they generally remain important only in a few product sectors – chiefly fruit and vegetables, and as suppliers to caterers and small independent retailers (including a wide range of non-food-specialist outlets such as garage forecourt shops). This trend is partly related to technological advances (in transportation systems and information technology, for instance) which seem bound to reinforce the trend. More importantly, it reflects the fact that retailers have taken over many intermediary functions, and wholesalers in many countries are finding their role restricted to the provision of services which retailers cannot provide or find it unprofitable to provide.

> **Functions of distributors**
>
> - Assembling a range of products from a variety of sources for offer to customers
> - Receiving bulk supplies and subdividing into smaller lots convenient for customers
> - Holding stocks in anticipation of customer needs
> - Providing additional services required by customers which also benefit suppliers (advice on use, promotion, credit, quality guarantees etc.)
> - Selling products and collecting and transferring payment to suppliers
> - Feedback of information to suppliers on product acceptability, level of sales etc.
>
> *Plus possibly:*
> - Deciding the customer selling price
> - Arranging for product manufacture

Table 5.1 Functions of distributors

This section therefore concentrates on the retail sector as 'channel captain', whose needs increasingly drive the entire system. A final section deals separately with the catering/hospitality ('out-of-home') sector, which accounts for a large and growing proportion of consumer food expenditure (Figure 5.3). The catering sector combines processing and distribution functions, and is strictly a 'food service' provider, but however it is classified, it represents an important customer for agricultural output which is worth targeting separately. It is also a very realistic target market for individual producers who can often supply the small volumes of locally-produced, innovative, quality products which many caterers require. Caterers are moreover generally easier for the small producer to supply, and are receptive to promotions of regional or sector produce, and their needs are sufficiently different from retailers to require separate consideration.

RETAIL DISTRIBUTION

Retailers are the marketing system's specialist salesmen, so it is not unreasonable that they should be channel captain in determining product and marketing specifications. They are in direct contact with the consumer, and this gives them unrivalled information about consumer requirements. They own the shelf space, and for most products they control access to household consumers. The revenue earned by the entire marketing system therefore depends substantially on the retailers' success or failure: if they increase their turnover it benefits their suppliers; if they fail in competition with other retailers, their suppliers lose as well as their shareholders.

This undoubtedly puts retailers in a position of power which they may be able to use to their own advantage. Without them, however, producers and processors would have to develop and invest in alternative distribution systems and acquire similar sales expertise – with no certainty of achieving equivalent efficiency or performance. Collaboration is therefore likely to be a more productive strategy.

MARKETING FOR FARM AND RURAL ENTERPRISE

Structure of the sector

Retail stores vary tremendously in size, in product range, in merchandising style and in employment generated. They are generally categorised (irrespective of size) by:

- product range: narrow-range shops like specialist butchers and greengrocers, or multi-product stores

- business organisation: independent traders, multiple chains, cooperative stores. (Independents range from one-shop businesses to symbol chains like Spar, which are composed of independents trading under a common logo and sourcing products through a central organisation.)

Whatever their size or product range, all retailers undertake a similar range of functions which serve consumers and the rest of the marketing channel, and the much-resented retailer's margin is the payment for these services. Since the 1970s the same broad trends have been evident in the retail store sector in most developed countries.

- *A decline in the total number of food retailers*, accompanied by a rapid fall in the number of specialist food retailers and an increase in the number of non-food retailers selling a more or less limited range of food items.

- *A rise in the market share of multiple firms* (more than ten outlets) at the expense of small independents. Most significant is the disproportionately large share taken by the largest multiples (some trading under a single name, others using different names). In the UK, for instance, five multiples have in excess of 50 per cent of total sales in all food outlets and 60 per cent of sales of all grocery outlets, and they are dominant in some product sectors (around 50 per cent of the milk supply).

- *A sharp increase in the range of goods stocked.* Food stores quickly added health, cosmetics and general household goods, and some moved into clothes and DIY. All new large stores have petrol stations and many have dry cleaners, banks, and pharmacies; most stock a wide range of quality fresh and delicatessen foods, and take-away meals.

- *A substantial change in the size and style of stores* to accommodate the increasing range of goods. The move from counter to self-service occurred in some countries in the 1950s, encouraging the development of supermarkets and the large-surface superstore or hypermarket. In the 1960s and 70s there was a similar move from a functional style of merchandising to increasingly attractive 'shopping environments' which have provided an ever-expanding range of customer facilities.

- *A dramatic development of retailers' own-label products* which has made retailers the channel captain in product innovation and development as well as distribution. Own-labels were initially part of a price-competitive strategy targeted at low-income shoppers, supplying a lower-price product which often reflected lower quality or packaging. They subsequently became part of the strategy to create a customer base loyal to the retailer rather than a manufacturer's brand, and most now combine quality with price competitiveness. Some multiple own-labels enjoy a status image of their own, and many are prominently advertised. (Figure 5.6 shows own-label penetration by product sector in the UK market, where the phenomenon is most advanced.)

- *A strong emphasis on price competitiveness.* Economic prosperity in the 1980s (and in the UK the high profitability of food retailing) shifted the emphasis from price competitiveness to store quality, product range, and customer service. Economic recession in the 1990s saw a return to price competitiveness led by cheap-and-cheerful multiple chains which offer a limited range of less well-known branded products with the minimum of services, at substantially discounted prices.

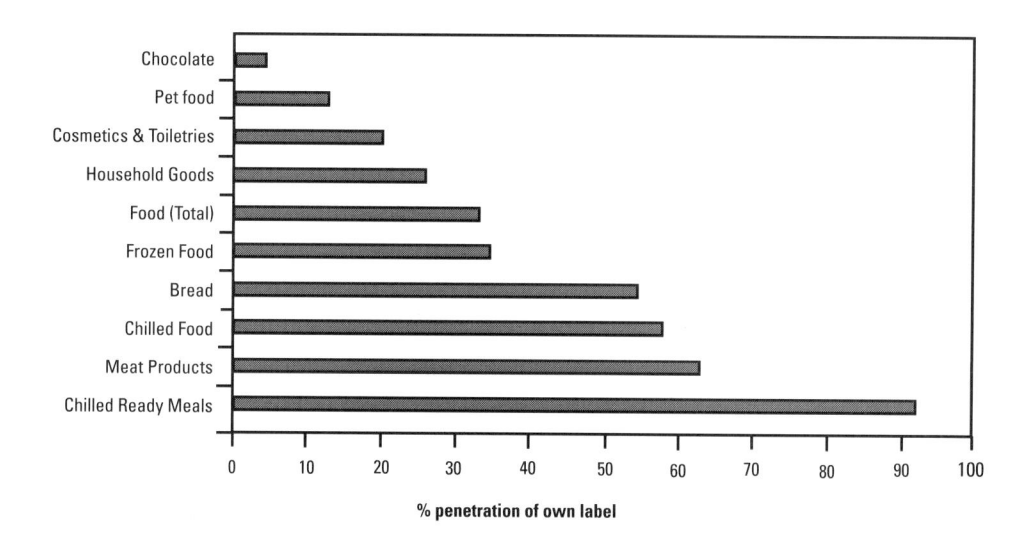

Figure 5.6 Own label penetration by broad product sectors (UK)

RETAILER MARKETING STRATEGIES

Retailer marketing strategies dictate the conditions of supply which any marketing offer must meet. They also provide a model of successful marketing management which is widely admired and emulated in other business sectors, and a practical illustration of the strategic considerations introduced in Chapter 3.

Like any other marketing manager, retailers can appeal to customers by manipulating three major interacting factors:

- prices and costs
- type and range of products sold
- customer image of the marketing proposition.

Prices

Reducing prices is an obvious sales strategy on which some retailers base their entire approach, but lower prices without lower costs lead to lower profits and the strategy is easily copied by competitors – particularly those with an efficiency or cost advantage of their own. For instance, the arrival in the UK of aggressively price-discounting foreign retailers in the early 1990s forced established discounters to cut prices even further, and compelled the quality multiples to become more price-competitive while yet maintaining their up-market image.

Some quality retailers responded by establishing a subsidiary multiple with a low-price trading philosophy and presentation, carefully differentiated from their up-market chain. Most retailers followed the high-risk strategy of resisting the price competition and stressing their up-market position, adding services as a differentiating characteristic to attract specific customer segments. This retained customers who valued the up-market offer, but the same customers also shopped at discount stores for heavily discounted, basic items. Most of the multiples therefore introduced customer loyalty cards which offer a discount or trading bonus, in the attempt to retain a loyal customer base whose repeat purchasing will maintain throughput and return on investment.

Costs

A price-led strategy depends on keeping down the primary cost of trading – the price of goods bought for resale – together with other trading costs: chiefly store-associated costs and logistic costs involved in distribution and stocking from production to retail point of sale.

Low-cost premises with minimal facilities are an obvious cost-cutting way of acquiring and retaining market share, but quality retailers resisted this. Instead, they concentrated on the search for ever larger efficiency gains in logistic costs, which they achieved by reducing the number of suppliers and cutting costs, or eliminating services whose costs are greater than the benefits received. This was made much easier by the introduction of direct product profitability accounting coupled with EPOS, which gives near-instantaneous sales recording, stock monitoring, repeat ordering and detailed customer information (products purchased, frequency of purchase, level of expenditure etc.) This improved internal efficiency levels and customer service, allowing even faster evaluation of new product sales and even faster de-listing of products which do not meet required profitability targets – with obvious implications for suppliers.

Retailer multiples also achieved further cost/price gains by undertaking manufacturer and wholesaler functions more cheaply and effectively, through own-label manufacture and contracted production or supply links with primary producers. Both give the retailer greater control over product specification, production and the resulting costs (inputs, management etc.), but they can also benefit primary producers.

Own-label manufacture maximises retailer control, and was resisted by manufacturers who had an established brand to defend (although most now undertake own-label production as a necessary evil). For new suppliers or suppliers without a strong brand of their own, producing retailer own-labels is a way of breaking into a market without investing heavily in the advertising necessary to establish and support a brand. It may also reduce new product development costs where the retailer either develops the product and simply orders its manufacture, or shares the cost with the producer.

Own-label manufacture also enables small producers to supply multiple retailers, for although the volume requirement of a multiple may bar a small supplier with his own branded product, several suppliers can meet the requirement by supplying a product under the retailer's label. Similarly, one supplier can manufacture for several retailers, putting different labels on different product runs. This provides a means of expanding production with less effort and cost than are required to build sales of a branded product. On the negative side, producing own-labels leaves a supplier exposed if he loses his buyer and has no brand of his own to fall back on. Own-labels also allow the retailer to engage in price competition by reducing manufacturing margins. The advantages of own-label manufacture/supply must therefore be weighed against the possibility of seeing margins squeezed by an all-powerful buyer.

Supply links and marketing alliances with primary producers can give retailers almost the same control over the product delivered as they achieve by own-label manufacture. These arrangements also benefit the retailer's cost-reduction strategy by reducing the marketing cost. They are consequently an important part of the retailer's drive for cost and management efficiency within the marketing channel.

From the producer's point of view, such alliances provide a secure outlet and a price known before delivery. Product development and production changes are often jointly undertaken, reducing both the associated risk and cost to the producer. Marketing responsibility invariably lies with the retailer or a producer group, again reducing the risk and cost to the individual as well as the marketing skills required. On the negative side, supply may have to be forward-contracted (Chapter 9), and production and marketing discipline is demanded by the retailer and any marketing group through which supply is organised (Chapter 12).

Product range

Once customers are in store it is vital that they buy as much as possible in that one store: consequently, product range (number and variety of products) is a prime consideration for the retailer. It is also consistent with the consumer's preference for one-stop shopping, which is so strong that most consumers will not go out of their way to buy a product, however good it is.

The range of products in most stores has consequently continued to expand, as retailers seek innovative products and services with which to tempt customers and maximise the return on their investment. The product mix varies with time and place, reflecting changing consumer demand as well as economic and cultural factors, but the twin objectives of this diversification are always increased profitability from higher-margin items and the ability to attract one-stop shoppers. Basic food purchases like milk, butter and bread are routinely used within this strategy as 'loss leaders', sold at subsidised prices to attract additional purchases and new customers (Chapter 8).

Running against this general trend, the 1990s have seen a revival of highly specialist retailers whose range of products is *deep* rather than *wide* (craftsmen butchers and bakers, quality cheesemongers), whose competitive strength is quality of product and expert service in a narrow product area. The emphasis in these outlets is strongly on innovative products (often revivals of traditional products such as regional cheeses and bakery products), together with superior quality of product and presentation (above all, lightly processed and 'hand-finished'). The quality multiples provide a similar offer via their serviced delicatessen and fresh products counters, and even some discount chains have in-store franchisees (typically a butcher or a greengrocer) whose service and product mix offset the store's standard, price-led offer.

The acquisition of EPOS data on customer purchases has allowed retailers to identify much more sharply the relative contributions of individual products and product sectors within a store's range, and thereby to exclude poor performers. In the foreseeable future the major food multiples are unlikely to use this enhanced power to restrict their product range, because supplying all their customers' needs is a major part of their strategy to achieve a loyal customer base committed to one-stop shopping. The pressure on individual products to perform well within the product mix will nevertheless continue to intensify.

Customer service and company/store image

Customer service and company/store image are leading components of retailers' competitive strategies. Image is a function of the products and services supplied, together with a corporate identity which is communicated through the entire customer offer, to convey a quality and value-for-money message which encourages repeat purchasing (Chapter 11). Considerable resources are devoted to identifying and modifying a company image to match an identified customer segment, and to the communication of this image throughout the entire chain. This means that any supplier's offer must be consistent with the company/store image – which explains the inflexible product and delivery specifications that suppliers have to meet.

The tendency over the last thirty years has been for all stores and chains to move up market, and to change their image appropriately. By the late 1980s this meant a generally high and continually improving level of service, which means that increasing ingenuity is necessary to differentiate one retailer's offer from another (for example, special provision for the elderly and disabled, or shoppers with children; privileged banking and financial services). Renewed price consciousness in the early 1990s also tested customer loyalty to particular stores and chains, and a combination of competitive price together with the maintenance of a strong store/company image is the long-run strategy of most major food retailers.

FUTURE TRENDS IN RETAILING

The retail food sector has undergone massive internal changes in the past 30 years, and in the process has transformed and raised consumer expectations of what constitutes an acceptable product. By the end of the 1990s some observers forecast a new revolution in retailing, prompted by changes in the marketing environment. The marketing proposition of the last thirty years has been to provide a pleasant shopping environment for the car-borne consumer, yet many countries are now imposing planning restrictions on out-of-town developments, and restrictions or penalties on car use. The potential for home shopping by computer has also led some observers to predict that multiple retailer supermarkets will become redundant dinosaurs stranded in their out-of-town complexes.

Past performance suggests that this judgement is premature, and that the large multiple retailers will continue to adapt their offer to fit the environment. Some have responded to planning restrictions by re-opening small high-street stores and narrow-range convenience outlets, trading in their own name or in joint ventures with other companies (for example, in petrol station outlets). Most multiples were quick to introduce direct marketing of some products (typically wine, flowers, gift food packs), and home delivery for orders placed by telephone or via the Internet.

Will multiples miss the home shopping boat?

'The supermarkets are sceptical about home shopping because they have invested in a complex and expensive infrastructure which they say is here to stay ...' This was the view of Lee Eskholme, executive director of Food Ferry, a London-based home delivery grocery service. 'The big five are going to have to reconsider their current investment plans ... It would be foolhardy to suggest supermarkets will die, but they will definitely have to shrink in numbers and store size and cut down on product numbers.' In research carried out ... recently, an astonishing 67% of those questioned said they were comfortable trading electronically ... The future of retailing, according to Eskholme ... is about partnerships between companies in the various sectors of the food and drink industry fronted by a major brand. 'Wholesalers, distribution firms, an electronic broker and village shops, c-stores and forecourts, are the perfect mix. The brand name would provide the element of trust consumers need and it would not necessarily be a food and drink name.

The Grocer, 20 July 1996

Box 5.1

The development of such initiatives is likely to depend on consumers and governments rather than on retailers, and their impact on suppliers cannot be predicted. What is certain is that retailers in a food market which is static in volume, and almost static in terms of real expenditure, can only grow by taking market share from each other or by further mergers which reduce the number of firms and stores. In either case there will be increased pressure to squeeze suppliers' margins, and further pressure on suppliers to deliver to buyer specification which will be difficult to resist if the total number of outlets available declines.

Suppliers may also have to accept a binding supply agreement as a necessary condition of trading, either negotiated directly with a retailer or through a marketing group. Such contracts will not necessarily bring a marketing premium, but they will invariably include production and marketing specifications and penalties for default. Suppliers may also have to tolerate longer payment periods than they are used to in traditional markets, as a condition of market access. They will certainly have to meet retailers' exacting delivery and stockholding requirements, and adapt their operations to those of the retailer, not vice versa. For example, retailers have already imposed on food manufacturers – including small processors – the responsibility for management of retailer stocks (*vendor-managed inventory*), which transfers a significant cost burden to the supplier. Suppliers can also expect retailers to be ruthless in shedding suppliers who cannot accept this discipline. On the other hand, suppliers who consistently deliver to these high standards can be more confident of the long-term sustainability of their markets, and will earn any marketing premiums which the system can deliver.

> ### *Taking loyalty to the door*
>
> Multiples are branching out with delivery services – another step forward in a bid to harness customer loyalty. Tesco launched a pioneering home-shopping service for web-site users from its Ealing store, distributing Tesco Direct catalogues which list 18,000 products ... Budgens announced plans to launch a home delivery service from its newest store ... As long as £20 is spent per shopping trip, a customer can receive delivery via a cold-storage van within a three-mile radius. The service operates six days a week ...Somerfield's store in Crieff, Scotland, has been running a similar operation for three years and five M & S outlets have been doing it for 18 months ...'Order and Collect' is Sainsbury's brainchild. A shopping advisor accompanies customers around the store to draw up a weekly list and the goods are then ordered by telephone and packed by the staff for a £2 charge.
>
> *The Grocer, 21 September 1996*

Box 5.2

CATERERS

By far the largest share of food consumption is accounted for by household purchases, but the share of the catering and hospitality sector is growing in most countries, and in some countries rapidly. The range of outlets is vast, but available data are conventionally grouped in two broad categories:

- the profit-making sector: includes a wide range of private firms providing an immense range of services and style (Table 5.2)
- the non-profit (or 'cost') sector whose objective was traditionally to provide a service rather than generate a profit (usually institutional or workplace catering, including educational establishments, hospitals, armed services, welfare catering).

The distinction between these two sectors has become less valid since the 1980s because many 'non-profit' organisations are now required to generate at least some margin from catering, but data will still be found recorded under these headings.

Both sectors are characterised by an immense variety of style and size of establishment. Institutional caterers range from factory food production lines serving thousands of meals, to gourmet kitchens which supply executive office suites. The profit sector ranges from the local take-away to the 5-star restaurant; from the owner-managed tea room to international fast-food chains with an outlet in every major town.

The cost sector has been in decline for two decades, but the profit sector continues to grow in most countries. In the UK profit sector, for instance, a ten per cent growth in meals was recorded between 1991-95, though this was not reflected as strongly in value, where an increase of six per cent was recorded. In any one week some 50 per cent of individuals buy meals outside the home. The age profile of the UK market is biased towards the under-34s and against the over-65s, a fact which is reflected in the highly differentiated growth rates of fast-food outlets. Another feature of the UK market is the rapid growth in the share of in-store supermarket cafeterias, which are generally serviced by the same suppliers – often with the same products – as the retail store. The overall sector shares are shown in Figure 5.7.

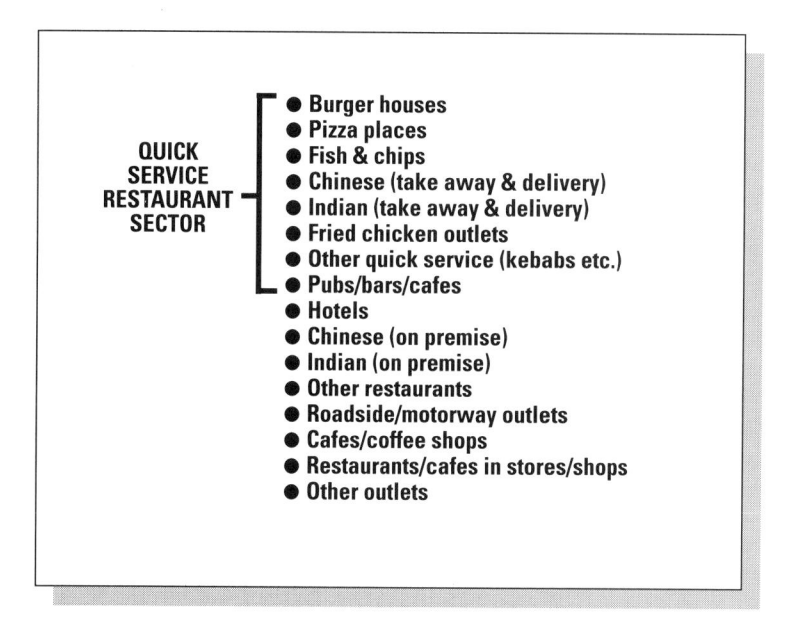

QUICK SERVICE RESTAURANT SECTOR
- Burger houses
- Pizza places
- Fish & chips
- Chinese (take away & delivery)
- Indian (take away & delivery)
- Fried chicken outlets
- Other quick service (kebabs etc.)
- Pubs/bars/cafes
- Hotels
- Chinese (on premise)
- Indian (on premise)
- Other restaurants
- Roadside/motorway outlets
- Cafes/coffee shops
- Restaurants/cafes in stores/shops
- Other outlets

Table 5.2 Profit catering sector

The USA has the largest, most highly developed and widely used out-of-home market, which may indicate the direction in which the UK, the rest of the EU and Australasia will follow. For example, the latest figures show the US per capita purchase of meals outside the home as 106 per annum, compared with 41 in the UK. The sector shares are also different: in 1994, for example, burgers and pizza represented 24 and 14 per cent respectively of the US market, but 10 and 18 per cent of the UK market. Pubs and restaurants had a 31 per cent share in the UK, compared with only 10 per cent in the USA. These figures reflect cultural differences which may never be completely eroded, but convergence is occurring everywhere; local opportunities for development potential may therefore be identifiable on the basis of published figures and trends.

The trends which are likely to affect the out-of-home sector are shown in

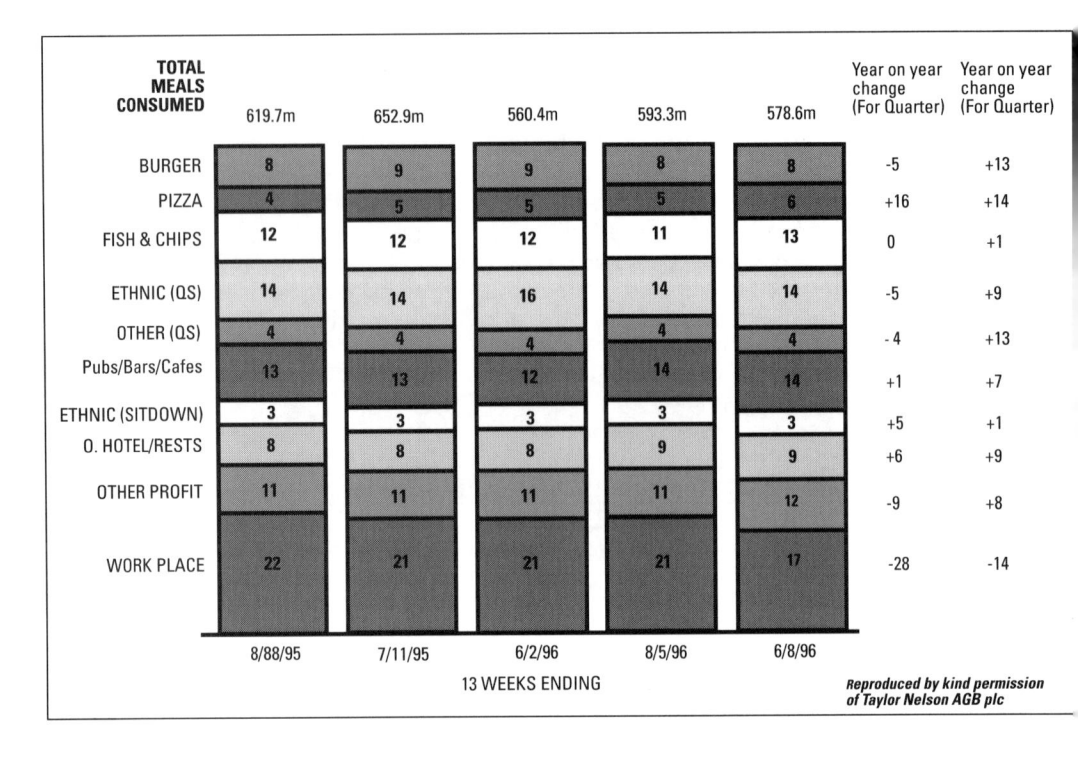

Figure 5.7 Catering sector shares

	8/88/95	7/11/95	6/2/96	8/5/96	6/8/96	Year on year change (For Quarter)	Year on year change (For Quarter)
TOTAL MEALS CONSUMED	619.7m	652.9m	560.4m	593.3m	578.6m		
BURGER	8	9	9	8	8	-5	+13
PIZZA	4	5	5	5	6	+16	+14
FISH & CHIPS	12	12	12	11	13	0	+1
ETHNIC (QS)	14	14	16	14	14	-5	+9
OTHER (QS)	4	4	4	4	4	-4	+13
Pubs/Bars/Cafes	13	13	12	14	14	+1	+7
ETHNIC (SITDOWN)	3	3	3	3	3	+5	+1
O. HOTEL/RESTS	8	8	8	9	9	+6	+9
OTHER PROFIT	11	11	11	11	12	-9	+8
WORK PLACE	22	21	21	21	17	-28	-14

13 WEEKS ENDING

Reproduced by kind permission of Taylor Nelson AGB plc

Table 5.3. The trend most relevant to a processor is the growth of fast foods, linked with 'de-skilling' of catering staff, which provides opportunities for expanded production of added-value, ready-prepared products. These products are likely to be required in large volumes, at attractive prices (particularly if the trend also continues for consumers to be more conscious of value for money); this market is therefore likely to be supplied by larger food manufacturers. However, there are many niches for genuine speciality products, especially farm-produced and regional products which allow a caterer to differentiate his offer, often in association with a hospitality or a 'Taste of Somewhere' scheme. Opportunities for the smaller producer also lie in the supply of fresh products to the up-market restaurant and hotel sector, where high-quality and specialist supplies are likely to remain at a premium.

Catering demand

The types of product used by the sector and the rates at which volumes are changing are influenced by the wide diversity of outlets, different geographical spread, and changes in the share of different sectors. All types of food are purchased, with approximately half the market represented by fresh or commodity products, dominated by fruit and vegetables, meat, poultry and fish. Customer preferences in the last decade have increased the share of fresh foods at the expense of processed inputs; at the same time, the use of complete ready-to-eat meals and prepared ingredients has expanded massively in proportion to the increased use of microwave ovens and the availability of acceptable microwavable products. Some outlets rely heavily on ready-prepared foods: this is particularly true of pubs and brewery-owned eating house chains, which account for some 50 per cent of all convenience products used in catering (compared with only 14 per cent used by hotels).

Catering use is particularly important for sales of some foods: for example, fish (16 per cent of total fish consumption), 'exotic' meats such as venison or rare breeds, and all kinds of speciality products (particularly gourmet portion-controlled dishes). Catering use of meat accounts for a significant proportion of total consumption, with most consumers still identifying meat as their first choice when eating out, although they rarely cook it at home. Promotional activity and product development work aimed at caterers has indeed been instrumental in sustaining total meat consumption, and in expanding sales of some meats which were not previously favoured (lamb, for instance).

Catering establishments play an important role in introducing new products to consumers. Ethnic foods are an obvious example, but venison, game birds and even trout and smoked salmon did not feature prominently in supermarket cabinets until their acceptance was first demonstrated in catering sales. The sector has also been instrumental in changing agricultural production via its specific product and variety requirements. McDonalds' insistence on the Burbank russet potato to supply its huge demand for French fries is the classic example. Less well known but arguably more important have been extensive changes in livestock breeding and husbandry systems, designed to supply products specifically developed for and promoted to caterers.

- Continued (slowing) growth
- (Even) less formal/everyday
- Growth of 'brands'
- Increased product diversification
- More sophisticated older consumers
- Move to non-traditional sites
- Development of consumer databases
- Increased emphasis on profitability and productivity
- Growth through disciplined menu areas
- New themes
- De-skilling
- Core businesses
- Operator monopolies, acquisitions and mergers and cross-branding
- Profit/cost sector merging
- High street multiples entering market

Table 5.3 Eating out: future trends

Different types of catering establishment obviously have very different requirements in terms of products, product quality, prices and supply arrangements. The institutional sector often produces written quality specifications against which suppliers are required to tender. Many fast-food restaurants are franchised, and inputs are frequently made and supplied to central specification in order to meet their price, volume, and consistency requirement; occasionally, the franchise owner may be a food manufacturer. It is therefore difficult for a small independent producer to access this market, but there is no reason in principle why a small producer cannot start a small-scale franchising operation to add value to farmhouse production: for instance, ice-cream parlours and milk bars.

All caterers without exception demand value for money, which is their particular combination of price, product and service. In the cost sector, some establishments define value for money with greater emphasis on price than on quality or consistency of product. For some up-market establishments, by contrast, product quality and exclusiveness may be required at almost any price; access to these is therefore valuable both financially and for the status it gives to the product. For the small, rural-based business with a high-quality or a special product, local independent hotels or restaurants are a realistic marketing target. The volumes likely to be sold and the costs incurred (in delivery, invoicing, after-sales servicing etc.) may be insufficient in themselves to be cost-effective and profitable, but the same is also true of servicing small independent retail stores. If catering outlets can be combined with store outlets, a supply operation which is only marginally viable may therefore become much more attractive.

Large hotel and restaurant chains (some of which supply the airline meal trade) may seem unlikely markets for the small supplier, but entry to the market has been achieved by small specialist suppliers of distinctive products. Single-portion farm-branded products are a simple way of adding value to a hotel beverage tray or a standard airline meal, and prestige hotels and international airlines are already serving farmhouse-produced and branded products such as biscuits, spring water, cheese, yogurt, and clotted cream. For the producer, such markets have a particularly valuable demonstration value, by introducing products to potential customers who would never normally see them, in a favoured situation where there is no exposure to competing products.

From the small producer's point of view large buyers also have the advantage that they are easy to supply once an agreement is reached, since they have centralised processing or distribution units. The problem for the small supplier is likely to be insufficient volume, even in a relatively local area or for a single outlet, so this may be a more feasible proposition for a producer group.

CONCLUSION

Access to raw materials – including non-agricultural substitutes – is not a problem for food processors and retailers: the only concern is their relative cost. Oversupply of processed products and retail services is also sharpening interfirm competition for market share and profitability, further intensifying the downward pressure on raw material prices. As agricultural price supports decline, this will intensify the pressure on farmers to improve the competitive position of their products and their presentation, through a better understanding of the shared objectives of the marketing system and more vertical coordination between producing, processing, retailing and food service interests.

CHAPTER 6

Analysing the market

Markets are typically composed of sub-markets with distinct product requirements; a business which can identify and supply such a demand may therefore gain a competitive advantage or a premium for a well-targeted product. Markets may also be segmented in the attempt to increase market share. This is achieved by identifying changes to the marketing mix which persuade existing customers to buy more and non-users to become customers.

This depends on the ability to identify exactly who the customer is, what requirements the product needs to supply, where, when, how, and at what price. It also requires an understanding of the way in which customers make purchasing decisions, since this allows the manager to influence the decision through the marketing mix. This is true whether the customer is a consumer, a processor, or an intermediary buyer purchasing for re-sale.

None of this is as intimidating as it sounds. Market research is neither the time-consuming exercise it is thought to be nor the preserve of large firms with big research budgets. Few special skills are required: it is largely a matter of data collection and analysis which all managers must undertake in the absence of a ready-made buyer 'spec'. The task has been greatly simplified by the development of good computer software which assists analysis (the only complex part), and the availability of complete market research packages. If more sophisticated analysis is required or a large investment is involved, research can be commissioned from a small research company. The cost need not be prohibitive since the research requirement for most small-scale developments will not be extensive, and the cost can be minimised by a manager who knows what questions need to be asked.

This chapter identifies the kind of questions which need to be asked, which managers in most circumstances will be able to answer for themselves. It begins with an outline of customer psychology, which many managers are less confident of analysing, though it is the richest potential source of added-value opportunities in most product sectors.

UNDERSTANDING THE CUSTOMER

A customer buying a product is not generally looking for a 'product' so much as the means to achieve objectives and benefits. For instance, a farmer who buys a tractor is purchasing the means to do a job, and the choice between alternative products is based on his estimation of:

- *functional product characteristics*: power, fuel consumption, stability on uneven ground, reliability, price etc.
- *added value*: good after-sales service, a friendly local dealer, a price discount – even a prestige brand name.

Consumers buying food products buy:

- *functional characteristics*: nutrition, a satisfying eating experience, price, taste, fat content, minimal waste, etc.

- *added value*: convenience in use, a preferred supermarket, price discounts, manufacturer brands, ethical concerns like animal welfare and the environment etc.

Purchasers of farmhouse accommodation or leisure activities buy:

- bed & breakfast, en suite provision, children's activities, catering facilities etc.
- peace and quiet, comfort, entertainment, excitement etc.

Effective customer research therefore needs to establish not just the obvious intrinsic product properties required for a given market/customer segment – this is the minimum that has to be supplied. It also needs to establish the other benefits sought by customers which are a source of endless *product augmentation* opportunities, and the circumstances in which the need is recognised and the purchase made. A producer who understands this may begin to earn some of that elusive added value so often talked about but rarely achieved.

THE PURCHASING DECISION

The decision to purchase a product is the result of a problem-solving exercise which starts with the recognition of a need, and proceeds via information search and analysis to make a choice between alternative offers (Figure 6.1). This process is evident for expensive and occasional *high-involvement* purchases like a tractor, car, house, holiday or consumer durables, where cost and inexperience prompt a systematic search for alternative solutions. Where routine *low-involvement* purchases are concerned (petrol, the weekly food shop), the decision-making process is less obvious but still motivates the selection. The process has become habitual: need recognition, evaluation and purchase are almost instantaneous under the influence of experience and familiar decision-making parameters such as known brands or a familiar store.

Whether the behaviour is high or low involvement, the decision can be influenced by changing the information input and the decision environment. To stimulate interest in a new product or a routine purchase which faces strong competition, all the components of the marketing mix are mobilised to attract and retain customer interest and purchase: distinctive or new product attributes, price incentives, optimal location, advertising and promotion. To sustain habitual purchase, *reinforcing activity* is necessary to ensure that the environment and the decision-making parameters do not change, so that product selection is almost automatic and the competition has to work harder to attract attention.

Habitual buying behaviour is more difficult to influence than occasional purchase precisely because it is almost automatic. This makes it difficult for a new supplier of a routine purchase to be considered as an alternative – especially if the customer has to make an effort to search out the product. A new product therefore needs active promotion to attract attention and minimise the customer effort and risk involved in trial purchase; continuing effort is then required to sustain awareness and gain repeat sales. A major marketing objective is consequently to convert a high-involvement into a low-involvement purchase, in order to reduce the marketing commitment necessary to attract attention (particularly the high promotion costs involved in maintaining product awareness in a competitive market).

ANALYSING THE MARKET

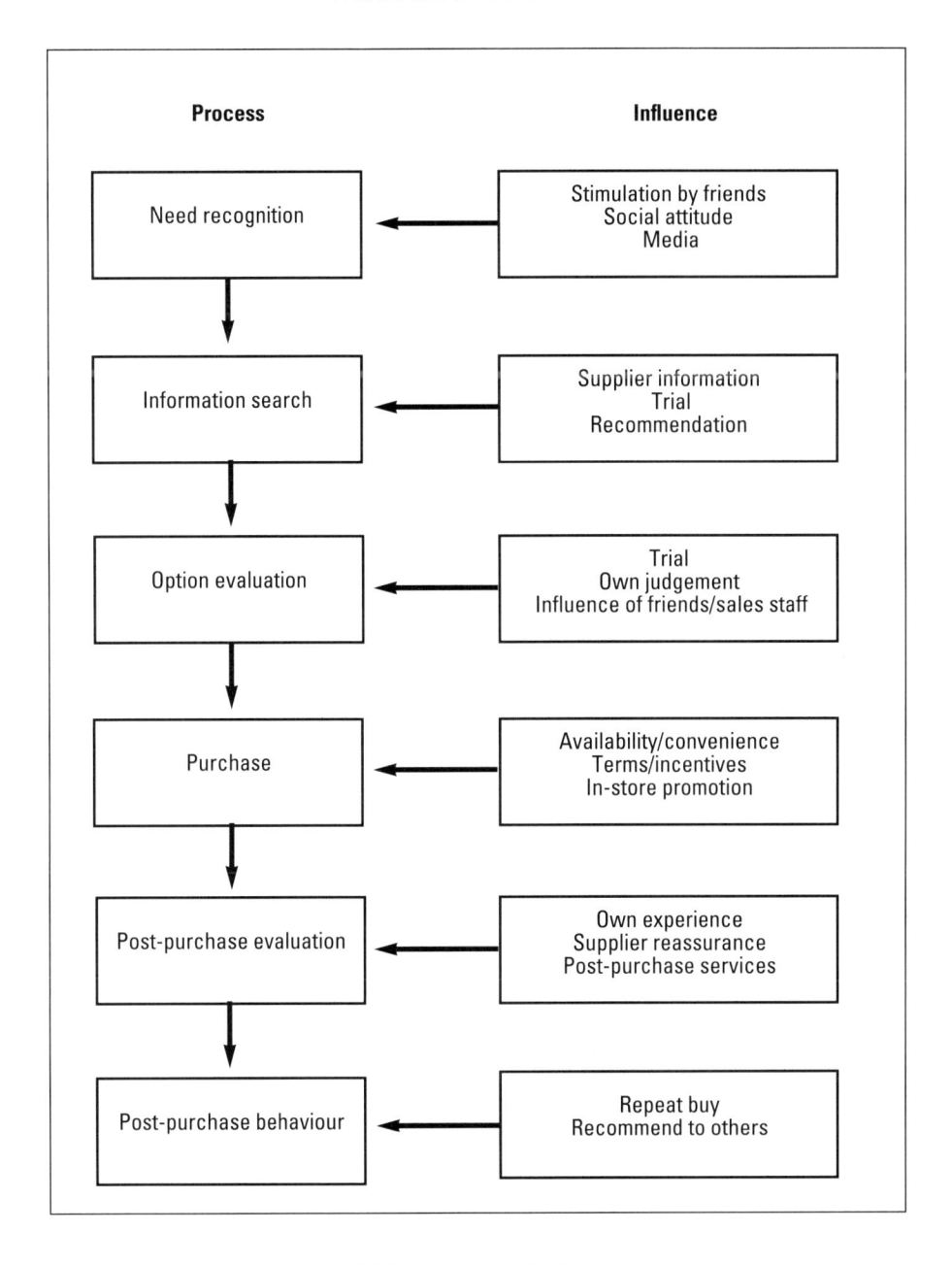

Process	Influence
Need recognition	Stimulation by friends / Social attitude / Media
Information search	Supplier information / Trial / Recommendation
Option evaluation	Trial / Own judgement / Influence of friends/sales staff
Purchase	Availability/convenience / Terms/incentives / In-store promotion
Post-purchase evaluation	Own experience / Supplier reassurance / Post-purchase services
Post-purchase behaviour	Repeat buy / Recommend to others

Figure 6.1 Consumer purchasing process

A notable feature of recent food marketing has been the success of the opposite strategy, which transforms low-involvement purchases into special items: for example, organic, farm-fresh, or environmentally-friendly food. A growing consumer awareness of food quality in general has also underpinned many quality marketing offers: 'luxury' yogurt, 'premium' beef, 'genuine all-butter' bakery products. However, need recognition and the decision-making process are still relatively informal for most food products, since they are bought week-in week-out by consumers who have little time for pre-purchase evaluation, and a strong predisposition to avoid risk by buying known products. The problem is therefore not so much to stimulate need recognition, as it might be for a new car, but the need for a particular product in competition with other foods.

In the case of food the competition is almost unlimited. Consumers rarely set out intending to buy lamb or beef, but a satisfying eating experience, and this need can be met by a vast range of alternative products which includes other meats (pork, poultry, venison), fish, and ready-prepared meals. The customer motivation which producers have to address is consequently not a search for lamb or beef of an intrinsic quality, but a search for added benefits like convenience, reliability, guaranteed production conditions and a clear customer conscience. There is no point in supplying this added value, however, unless consumers can recognise the product for repeat purchase – hence the need for product identification and branding.

Consumers and industrial buyers

Textbooks normally distinguish between consumers and industrial buyers on the grounds that the latter are more objective in their purchasing behaviour. Most industrial buyers (including farmers) would also like to think that this is so, claiming that their purchasing decisions reflect strictly functional production/manufacturing needs, management objectives, and operational and pricing objectives. However, industrial buyers, including farmers, have been shown to be as susceptible to a brand name or effective promotion as the average consumer. Where they act as principal on their own account (a retailer, a dealer in the mart, a wholesale merchant) the decision-making process is essentially the same as that of the consumer. For company buyers the process may be more formalised and laborious because other management layers are involved, and the individual buyer may have only limited (or no) freedom to make purchasing decisions. Identical processes are involved, however, and it is important neither to *over*-estimate the objectivity of professional buyers nor *under*-estimate the rationality of consumers, since both can be influenced via the marketing mix.

If any distinction is justifiable, it is with reference to involvement level, since industrial buyers are effectively in the same position as the consumer making a high-involvement purchase, when the search and evaluation stages are more extended and more thoroughly undertaken (Figure 6.2). However, a major objective of many professional buyers is to reduce the management time and cost involved in procurement, and the risk involved where regular supplies of consistent quality are required for a manufacturing process. This explains their preference for long-run trading relationships (*relationship marketing*), which reduce a high-involvement to a low-involvement purchase. Relationship marketing also brings parallel gains for the supplier, in the form of reduced marketing effort.

Who is the customer?

At consumer or industrial level, it is important to distinguish between the person who is making the purchase and the decision maker, who may well not be the same person. Where this is the case, it is the decision maker at whom the marketing offer should be targeted. This is obviously true of company buyers, who normally work to a specification and corporate policies determined higher up the management structure; company buyers are also not immune to the whims of senior executives. In dealing with large firms it may therefore be

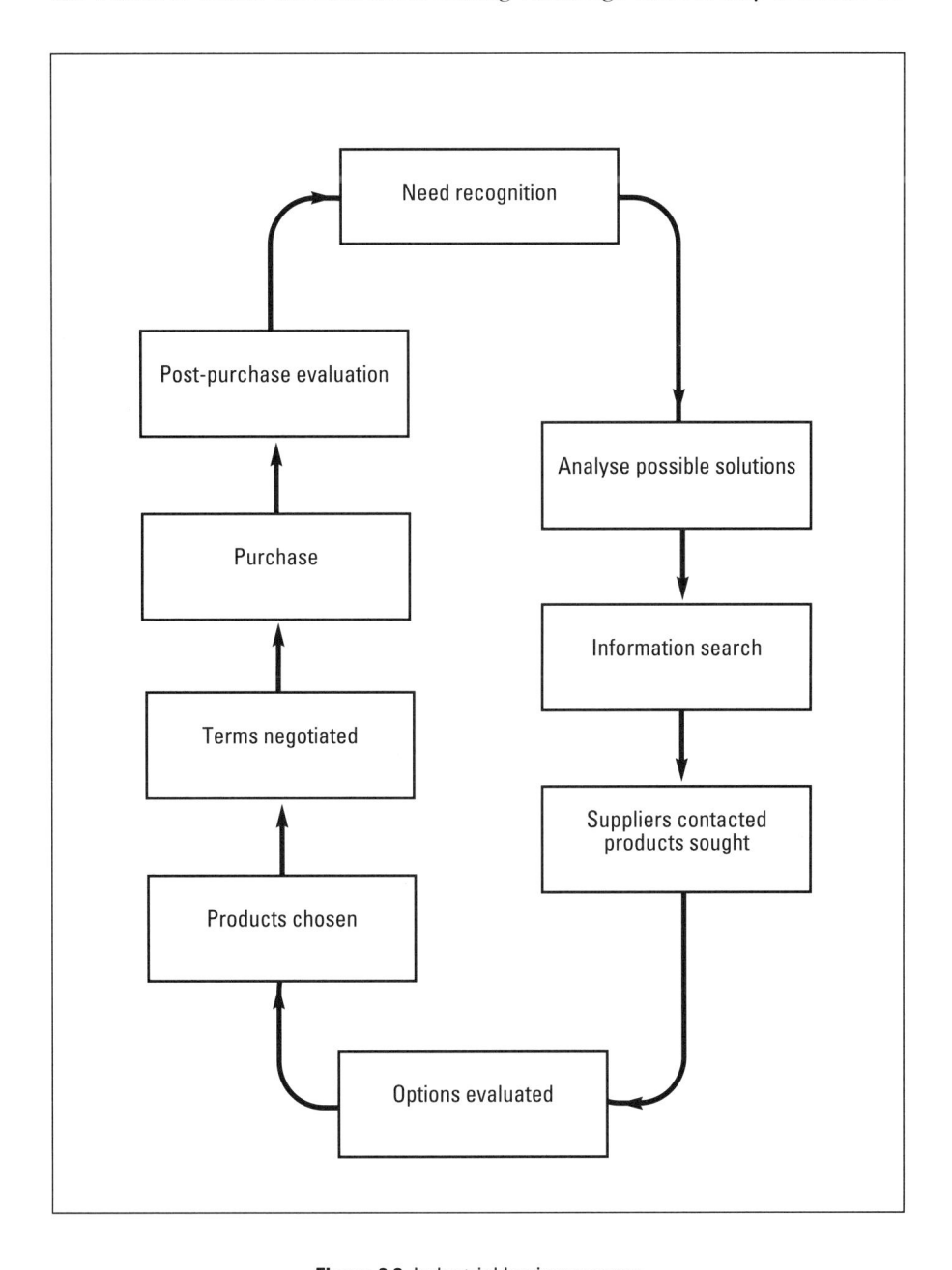

Figure 6.2 Industrial buying process

difficult and time-consuming to decide the right approach because of the management layers involved. However, a similar problem may also be encountered in the small retailer sector, which is characterised by family businesses with several decision-makers.

In the household situation children and partners exercise the same control over product purchases made on their behalf. The 'pester power' of children is especially known to motivate many food purchases by parents: the growth of vegetarianism and green consumerism, for example, has been strongly linked to the influence of children and teenagers. This is a boon to marketers, since parents with strong habitual purchasing behaviour will be influenced by their children although they may be impervious to other inducements. Advertising or product re-positioning targeted at children is therefore widely used to transform a parent's habitual purchase of everyday products into a high-involvement 'craze', and media advertising and in-store displays are frequently aimed at children. Pester power is also influential in larger purchases like family holidays, consumer goods and services, and this needs to be built into the marketing mix. In marketing farmhouse accommodation, for example, both facilities and promotion should target the demands of the young consumer as well as the product attributes attractive to their parents.

> ## *No meat and two veg*
>
> According to the latest Realeat survey ... sales of prepared vegetarian food grew by over 90% last year ... and in an NOP survey last February, more than half those questioned reported eating more meat-free meals ... Up to 25% of respondents identified a household or family member as a vegetarian, while 25% believed their children were likely to become vegetarian ... Children are increasingly making their voices heard, and they have fairly firm opinions.
>
> *The Grocer, 11 January 1997*

Box 6.1

ANALYSING BUYER BEHAVIOUR

Buyer behaviour has been categorised in many different ways. The simplest kind of model groups customers by personality characteristics and attitudes, innate and learned, which condition their reaction to products and marketing activity (Figure 6.3). These are affected by external factors including economic variables (price, payment terms), and socio-cultural influences which may reinforce or challenge other attitudinal variables (social class, demographic factors, age profile of households, religion, ethical values). All of these may exercise some constraint on purchasing behaviour. Alternatively, they represent 'special interests' which a well-judged marketing initiative can exploit to encourage sales.

Figure 6.4 shows a more detailed model which underlines the multiplicity of variables involved in consumer food purchases – which may suggest that every consumer may reasonably regard himself as unique. In fact, many consumers respond and behave in a similar way to one or several factors, and it may be possible to identify *reference groups* which share reasonably homogeneous characteristics. The object of market segmentation is to identify these groups and the customer characteristics which allow an appropriate marketing mix to be designed.

Reference groups may be large, or they may be tiny. They may be characterised by a single characteristic like geographical location (the number of households in a neighbourhood requiring doorstep delivery) or income (top earners). They are more likely to have a set of characteristics commonly referred to as *lifestyle*, each indicating targetable aspects of the marketing mix. A typical reference group might be: professional young families, middle class and reasonably well off, environmentally concerned, living in an up-market, semi-rural situation, who might become customers of a farm shop with an organic or environmentally-friendly product range. In researching their needs, other unknown product attributes or services may be discovered which would enhance the offer, for which they might be prepared to pay: for example, a coffee shop or a car-wash.

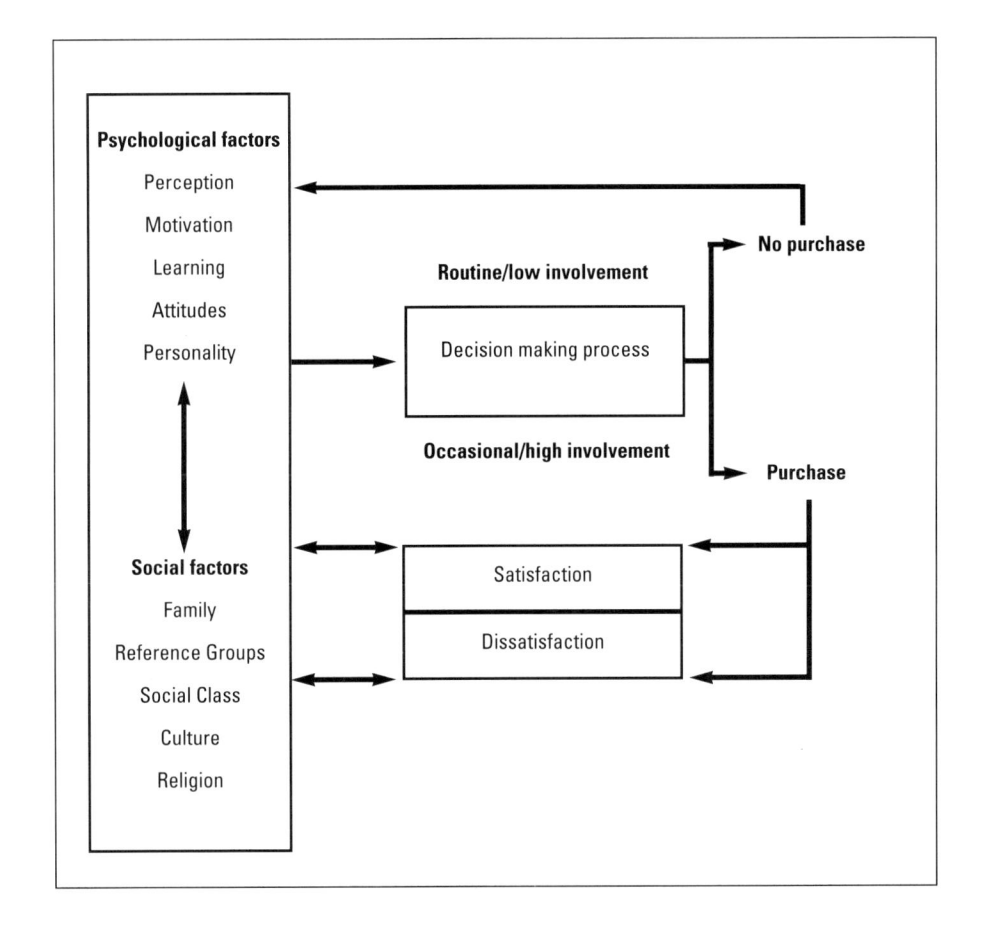

Figure 6.3 Understanding buyer behaviour

Within these reference groups opinion leaders ('trend setters') may be identified and targeted with promotional material, free samples or product presentations, and encouraged to try a new product in the hope that their favourable opinion will influence other buyers. (Opinion leaders for a rural business

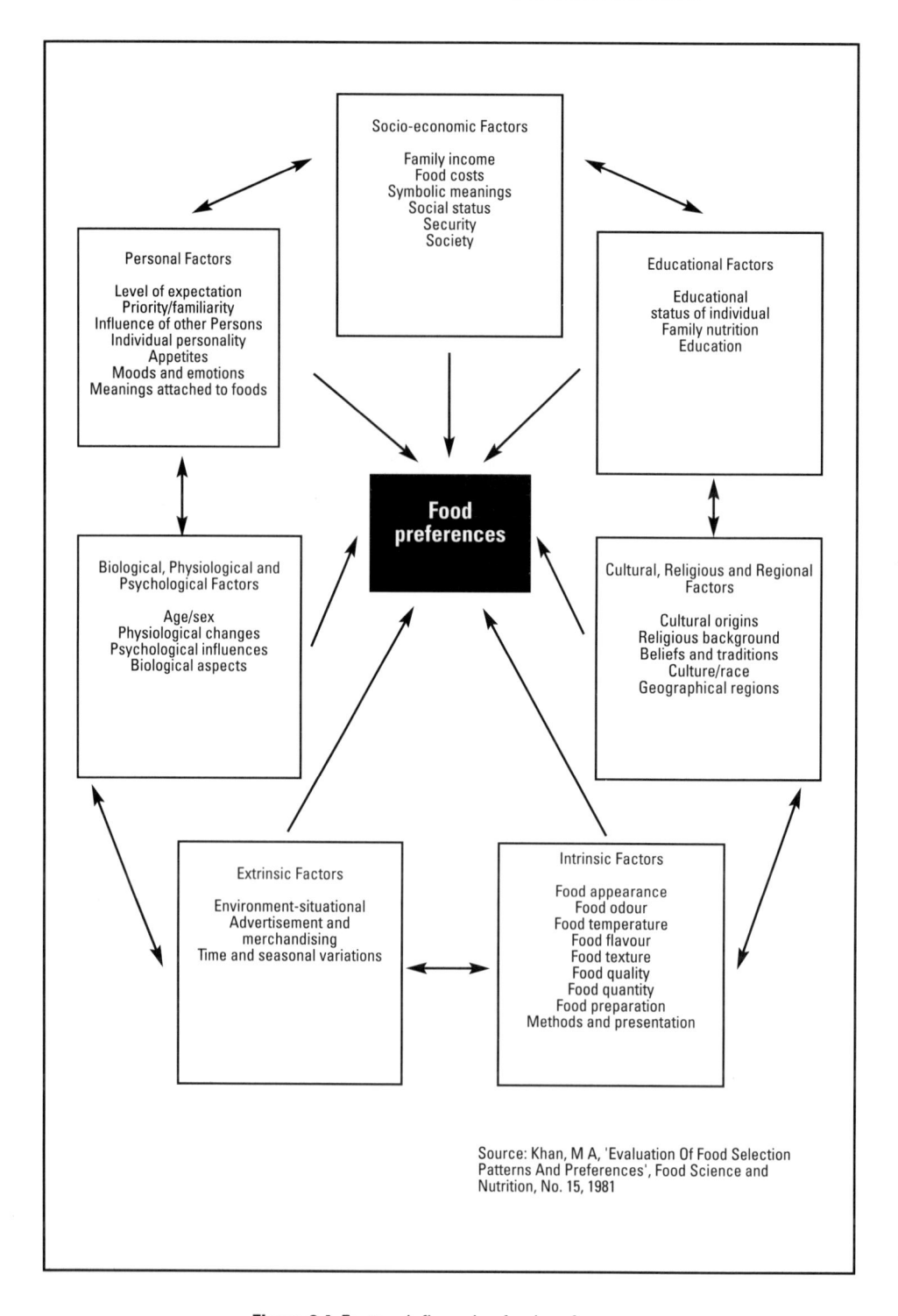

Figure 6.4 Factors influencing food preferences

could be WI members, the FE cookery class, a local school.) Attitude groups of customers may be particularly worth identifying: for example, environmentally concerned; animal welfare interest; the consumer group willing and able to pay a premium for organic farm produce.

Attitude group research is difficult, and is usually undertaken for commercial clients by market research companies. Some information usually trickles down to the professional and trade media, but it must always be treated with caution because it may be too client-specific to be useful. More importantly, expressions of intent, especially moral intent, are frequently not reliably converted into sales of product. A producer considering an attitude-based marketing proposition (organic produce, for instance) must therefore undertake original research to try to establish the number and commitment of realisable customers, their attitude to a price premium, and their precise product requirements.

> ### Price keeps organic meat in niche status
>
> Demand for organic meat is outstripping supply, but producers still have a number of hurdles to jump before this sector moves out of niche market status, according to industry leaders.
>
> Sainsbury's fresh meat senior manager, Tony Sullivan, said that 'The majority of customers are prepared to be green and environmentally friendly but only to a point, and that point is price. The premium for organic meat is between 10% and 15% higher than standard. Some shoppers are prepared to pay this difference but most would think twice about value for money.'
>
> The livestock procurement manager for Sainsbury supplier Lloyd Maunder said the product also had to sell on eating experience as well as being authentically organic. He predicted the current retail price premium would continue for the next five years, but would then reduce as market share started creeping up to 2%, 3% or 4%. At present, Sainsbury's weekly volume of organic beef is eight to ten cattle, which is 0.15% of total beef sales. This represents an increase of 500% on last year [before BSE]. For sheep the figure is 60 per week, 0.10% of all lamb sales. This represents no change on last year.
>
> *The Grocer, 11 January 1997*

Box 6.2

MARKET RESEARCH

The need for market research should be obvious in farm businesses which market their output direct to the consumer (and there is a surprisingly high number of these, even though only a small proportion of total farm output is marketed direct). The research should also be easy since they are in direct contact with their customers, yet the evidence is that their customer research is often rudimentary, and not designed to maximise the opportunity to obtain the most effective, competitive marketing specification. Research also tends to be undertaken only when a new product or market is under consideration, although it should be a continuing management discipline because it provides the information necessary to make decisions about *existing* products.

The need for market research by producers of industrial raw materials is rarely recognised as requiring more than a few phone calls to known buyers or the choice between two or three local marts. As Chapter 5 noted, however, many intermediaries and some marketing channels are poor communicators of end-use requirements. Many are also slow to identify market changes and new product opportunities, and they may hinder efforts to establish traceability, which is becoming a condition of market access. A knowledge of consumer as well as intermediary markets is therefore necessary simply to judge whether an intermediary is in the right channels and is an effective marketing partner.

IDENTIFYING MARKETS

Many marketing opportunities consist simply in identifying a new buyer for an existing product. An initial screening should therefore identify *all* potential users and product uses, starting with broad categories (processors, caterers, wholesalers, retailers, non-food manufacturers, consumers) and only then narrowing the search to markets worth detailed investigation.

The search should not be confined to the highest-value markets; processing outlets for lower-quality output or a seasonal production surplus can make a vital contribution to overall profitability. Similarly, although it is true that a high-quality product targeted at a quality niche market is generally a sound differentiating strategy for a small business with relatively low volume, industrial-use outlets are normally required for at least some of the output, and may themselves be targeted by market-building activity. For example, meat as an ingredient of ready-processed meals is a substantial industrial market which is capable of being 'grown', since the proportion of meat used in ready meals ranges from negligible to as much 33 per cent. The development of new products and new uses for existing products also needs investigation for its capacity to increase total consumption or create new markets: for example, lamb or pork in manufactured meat products in place of beef (Box 6.3).

Making the most of lamb mince

A rapid rise in sales of lamb mince has introduced many new customers to this versatile product ... The Meat & Livestock Commission has been very active in promoting lamb mince, both directly through retailers and also through consumer education initiatives ... Thus lamb has featured prominently in recipe leaflets for in-store distribution and also in PR and other support activities ...Part of the reason for lamb mince's success has been wider distribution and regular in-store availability. If shoppers know the product will be available they can make it part of weekly meal planning. Giving recipe usage suggestions at point of sale completes this loop ... There has also been increasing consumer interest in kitchen-ready or further prepared minced lamb products, following a trend where shoppers frequently seek 'meal solutions' rather than just 'ingredients' ... This is reflected by the importance retailers and manufacturers put behind new product development with British Lamb.

The Grocer 24 August 1996

Box 6.3

This is an area in which collective producer action is generally required, and there are abundant examples of marketing groups and sector-wide organisations which have successfully expanded demand by commodity-linked R & D

(Box 6.4) and generic promotion (Plate 10). Producers rarely recognise the value of this market development work by producer-funded bodies, and even more rarely see the use of their levy contributions to fund this activity as a deferred marketing cost which finds real markets for their output. Producer groups are also active in developing export markets, and have sustained demand in some sectors by delivering a superior product in controlled marketing conditions (Plates 1 and 9).

The potential to identify additional markets for an existing product is equally relevant to non-agricultural enterprises such as farmhouse accommodation and farm food processing, where the utilisation of installed equipment and employed labour needs to be improved by evening out seasonal imbalances in demand. In the right location farmhouse accommodation may be marketable for year-round business and conference use as well as for holiday lets. Farm restaurants can extend their highly seasonal business by catering for local weddings, baking for local shops, providing office lunch packs and boardroom lunches. Farmhouse processors have created new products to fill seasonal gaps: Christmas pudding ice-cream and yogurt; speciality catering products for a period of low sales and turnover but high consumer expenditure.

A business will often be able to identify such marketing opportunities from experience or observation. A new market may be suggested in the course of normal trading, by another producer or a trader in the mart, or by customer requests for product requirements not currently supplied. There will be situations, however, where a producer has no idea whether a market exists for a product or a product concept. This is most likely in:

- new geographical (especially export) markets (for instance, is there a market for lamb in Denmark?)
- markets outside the core business (was there a market for farmhouse accommodation, farm-processed cheese?)
- new product development (was there a market for bunkhouse accommodation, hedge-laying, habitat maintenance?)

Where a product already exists or there is a fairly firm product concept, potential customers may be inferred from product characteristics and uses. This is

---

HGCA enterprise awards target new markets for grain

Cereal farmers in the UK have laid down a £300,000 challenge to food manufacturers in a drive to find new markets for their wheat, barley, oats and rye. The Home Grown Cereals Authority (HGCA) Enterprise Awards, on offer to food and drink manufacturers with interesting plans to increase their use of home-grown grain, can be worth as much as £30,000 to businesses who make a successful bid to the fund. The awards' main sponsor is the HGCA, together with Food From Britain and the Meat & Livestock Commission. A vast range of products is covered by the awards, including bakery goods and food ingredients. Virtually any project which aims to increase sales of cereal-based food and drink is eligible ... The 1996 fund of £260,000 provided funding to develop 16 projects which included plans to develop a wheat-based rice substitute, convenience products and speciality sausages.

Food Ingredients & Analysis Journal, Nov/Dec 1996

Box 6.4

done by creating a product-by-use matrix, where use = product attributes, and product includes all potential competitors (direct competition from other producers, competing products in the same class or category, and competing product classes). A simplified example is shown in Figure 6.5; a real example would include much more detailed information about branded products.

This matrix will show if the proposed market is highly *congested* or if there is no competition. If it is congested, the question to be answered is:

- whether a unique selling proposition can be identified which will give a competitive advantage sufficient to out-perform the competition (by added value, product assurance, branding, etc.)
- whether a product can be re-positioned to appeal to a less congested market area, with or without modifications.

The apparent absence of competition in a market should not be taken as indicating a genuine opportunity to satisfy an unsatisfied need: it may indicate that the need does not exist. If the need does not exist it may be possible to stimulate it, as the food industry has demonstrated by creating demand for whole new classes of product within a market. Yogurt is the classic example, which was almost unknown in some countries twenty years ago, but is now a daily staple of the diet. Export demand for products may similarly be created by introducing into a country products unknown to its consumers. Yogurt again illustrates the point, since it was exporters, not domestic suppliers, who stim-

Product	Main Meal Home	Main Meal Rest.	Hot Snack	Cold Snack	Sandwich	Take Away
Roasting Beef	✓	✓	✓	✓	✓	-
Stewing Beef	✓	✓	-	-	-	✓
Leg of Lamb	✓	✓	✓	-	?	-
Stewing Lamb	✓	✓	-	-	-	?
Roast Pork	✓	✓	✓	✓	✓	?
Venison	?	✓	✓	?	?	?
Chicken	✓	✓	✓	✓	✓	✓

(Uses)

Figure 6.5 Product/Use matrix

ulated the new demand. However, both strategies require a sustained marketing programme, including product development and a strong promotional component: they are, in other words, expensive.

RESEARCHING THE MARKET

Once a potential market is identified it must be described as fully as possible in qualitative and in quantitative terms. The more detailed and accurate this description is, the more likely it is that the marketing mix will fulfil the requirements of a chosen segment, and the more reliable will be any estimate of enterprise viability. For instance, marketing drives targeted at elderly consumers have been known to get everything right with the product, but overlook the fact that the target customers have difficulty in opening packages and reading small print, may be confined to their homes or unable to visit large supermarkets, and unable to afford newspapers and magazines in which the product is advertised.

Market research must therefore identify customers both willing and *able* to buy, since willing customers do not constitute a market unless they are also able to buy – which means that they have the necessary money and the authority, and access to the product. Children often wish to buy a product but are not allowed to. Elderly or other income-restricted consumers may not be able to afford it. Others may have the money and the authority to buy, but do not shop where a product is available.

Market research falls into two broad categories (Figure 6.6): desk research, based on published secondary data (collected for another purpose) and primary (field) research designed to generate data specific to the marketing proposition.

The amount of information required and its degree of detail will vary with the business and the product offer. However, every business needs a basic minimum of information in order to draw up a production and marketing specification and to estimate sales revenue and costs. This allows a projected profit figure to be calculated, on the basis of which it will be possible to decide whether the market should be targeted. Much more detailed information is required as the degree of segmentation increases, since the risk increases when producing for a niche market. A trial marketing may sometimes substitute for detailed research by actually testing the market, and this hands-on approach appeals to practically-minded managers. A poorly conducted trial may produce misleading results, however, so some market research is always necessary. If the market requirements are very specific a professional input is advisable.

Secondary data

Information about *intermediary* buyers is difficult to obtain. Some data about public companies may be obtained from their Annual Reports, but information must generally be sought from press and media reports (trade and general) and sector or industry data published by government departments, development agencies, and trade associations. This gives a general idea of comparative performance, but the information is often difficult to interpret and to relate to the supply decision, which is why an understanding of consumer markets is also necessary.

Information about *consumer* demand for most products can usually be acquired, though with varying degrees of difficulty, detail, and cost. A surpris-

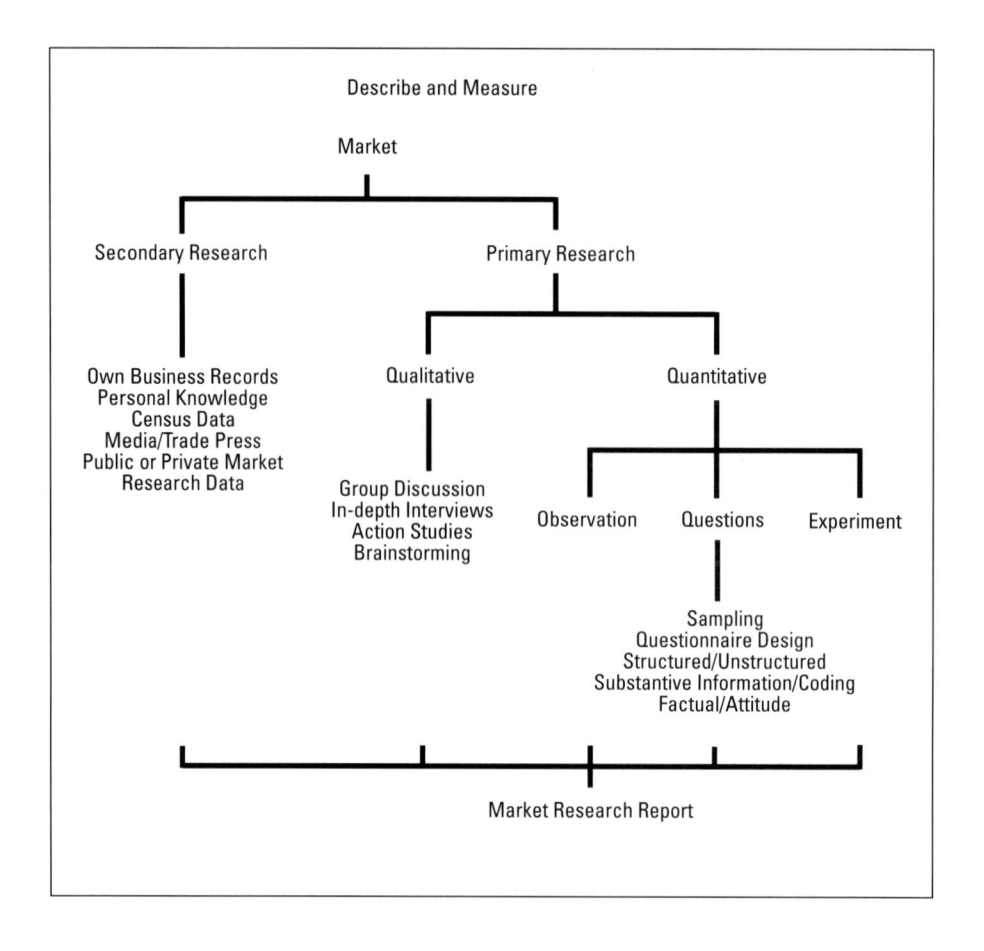

```
                    Describe and Measure

                          Market
                            │
           ┌────────────────┴────────────────┐
    Secondary Research                 Primary Research
           │                                  │
           │                        ┌─────────┴─────────┐
   Own Business Records       Qualitative          Quantitative
   Personal Knowledge              │                    │
   Census Data                     │           ┌────────┼────────┐
   Media/Trade Press               │       Observation Questions Experiment
   Public or Private Market  Group Discussion              │
   Research Data            In-depth Interviews            │
                            Action Studies                 │
                            Brainstorming            Sampling
                                                     Questionnaire Design
                                                     Structured/Unstructured
                                                     Substantive Information/Coding
                                                     Factual/Attitude

                           Market Research Report
```

Figure 6.6 Market research methods

ing amount is available free from published sources, and commercially published market research may be accessible free of charge through business support organisations, public libraries and colleges. A wide range of organisations also provide specific information free or for a small fee. The principal difficulty is to ensure that the information acquired is comprehensive and representative, and its interpretation requires particular care – a point illustrated below with reference to food consumption data.

Food consumption data

Food consumption data are published in most countries (for example, the annual UK National Food Survey and its equivalents in all developed countries). These are useful in indicating broad consumption trends, in total and between sectors, and they confirm that food purchases in many developed countries are virtually static, with a decline in traditional staples. In the UK, real expenditure on some items (adjusted for inflation) has actually fallen, and for some items where real expenditure has risen (for example, red meat) the proportion of total food expenditure they represent is constant. Others have seen an increase both in expenditure and in proportionate terms (fruit, fresh vegetables, poultry). The figures also confirm that consumer expenditure on added services (processing, distribution and marketing) has consistently increased as a proportion of total food expenditure.

The data allow regional and seasonal consumption patterns to be identified, and demand is usually broken down by demographic and socio-economic groups. The data are generally category-based, however (beef, milk, yogurt), and may exclude consumption outside the home, which is very important in some sectors. Statistics are also invariably out of date by the time they are published. Up-to-date market information about products in the market, manufacturers, retail sales figures, market shares, etc., must be obtained from market information services. Some of this is provided by producer organisations (for instance, MLC for the UK meat sector), but most of it is obtained by market research companies. For example, Taylor Nelson AGB produce the *Out of Home Eating Monitor* and weekly *Superpanel* data which give detailed consumption figures for all food products sold in the UK, together with brand shares, retailer shares, price etc. Few small businesses could afford to subscribe to these market research services, but the data (or abstracts) are often accessible via institutional subscribers like public and university libraries and business support agencies. The data are also available to other market researchers, which is one cost-effective reason for commissioning research.

Extreme care is essential in interpreting this information. The UK statistics, for instance, show that the Scots consistently consume less fresh fruit and vegetables than other regions, but this may be because supplies of the right kind are not available, or because the Scots dislike fruit and vegetables. Whether or not this presents a marketing opportunity, and what kind of marketing opportunity it might be, must therefore be thoroughly investigated. Similarly, published data show that single households purchase more fresh fruit and convenience foods than other groups. Before targeting them with single-portion packs, however, a farm shop would need to investigate its own customer profile, likely sales, and the cost and feasibility of adding this value.

Statistics are also limited in what they can tell you. For instance, they all state how much of an *available* product is *purchased*; they do not state how much is *consumed*, nor whether customers would prefer to purchase another product which is not available. Close study may also reveal that an apparent decline in one area reflects a transfer of consumption to another. In the meat sector, for instance, total consumption has increased slightly, but white meats (poultry, pork) increased at the expense of red meat (beef and lamb). Lower household purchases in most sectors reflect transferred consumption to the processing sector, as consumers substitute processed for fresh products. The loss is thus not absolute, and the transfer provides new supply opportunities for existing and new products. The shift of household demand to out-of-home consumption has already been noted, but it is also relevant that certain foods are only or mainly consumed outside the home. For example, consumers have always been wary of cooking game and exotic meats, and this lack of confidence extends increasingly to red meat, with the result that catering use accounts for a growing proportion of meat as well as game sales.

Consumer data

Information about consumers is more difficult to acquire from secondary sources because it is rarely published (as it is the result of private research which companies are obviously not keen to release to competitors). When information trickles down into the public domain it is usually market-specific and cannot be reliably transferred to another market, so consumer data must normally be collected through original research, or purchased from research companies.

A major development in the availability and detail of consumer data came with the growth of EPOS scanning in retail stores linked to individual customer information based on loyalty cards. This identifies precisely what every customer buys week in week out, allowing a previously unimagined degree of market segmentation – the result of which is visible in the increasing volume of individually targeted 'junk mail'. This information makes it theoretically possible to target individual consumers of (say) a farmhouse yogurt or cheese with promotional offers and price discounts, thereby maximising the impact of promotion and turning 'junk mail' into a useful customer service. However, although retailers have sold some of this information to commercial customers it is not clear whether they will release it to their suppliers, since this could also in theory allow suppliers to by-pass the retailer to sell direct to the customer.

Primary Data Collection

Original research is required to obtain both qualitative and quantitative data. Qualitative research generally precedes quantitative research, and is used to generate the concepts which need to be investigated, the questions to be asked and product ideas to be tested. It gives insights into perceptions, motivations and attitudes which cannot be quantified precisely, but which are critical to the actual number of products that will be sold, production costs, etc. The objective is therefore to attach values as far as possible to these qualitative factors, so that costs and revenue can be predicted. The three basic ways of collecting primary data are observation, experimentation, and customer surveys.

Observation is the classic way of obtaining information, and its importance is growing with the spread of IT and data interchange. Much of the secondary data that is available is generated through observation by professional market researchers, and will consequently be of a high standard and very accurate (for example, sales data from retailers, advertising monitoring figures, consumer-use diaries). The individual business will have similar information for its own existing customers, and from both sets of data it is possible to make projections about future sales, transfer of demand from one sector to another, change in shopping patterns, holiday locations, etc.

Experimentation is used to test-market products (by in-store and doorstep sampling, for example) and marketing methods (advertising and physical distribution trial runs). It may also be used to validate observation data. The statistics of experimental design and data analysis are now readily available via computer packages. Another form of experimentation, albeit less rigorous, is the action study in which new ideas can be tried out and customer reaction observed (for example, product-using sessions, test shopping).

Customer surveys are the principal tool of market research, carried out through personal interviews, focus groups and the structured questionnaire. The questionnaire survey allows greater quantification of results, but the quality of the output depends critically on the sampling methodology, questionnaire design and administration, and data analysis.

Since it is impossible to question all potential customers, a sample must be identified which is representative of the potential market, and the sample size is quite critical. Standard statistical methods are used which cannot be described here, but are available in specialist textbooks and software packages. If a large sample is considered necessary and the results are critical to business viability, the exercise is better contracted out to a market research company.

Whether the exercise is done by the business itself or contracted out, the questions to be asked need to be specified. The typical questionnaire contains *substantive* questions about the research topic (do you drink yak's milk?, where do you buy it, how much do you drink?) and *coding* questions which provide information about the respondents which will then be used for market segmentation (income group, age, sex, lifestyle, etc.). Questions may be *open* and/or *closed*. The latter restrict the response to a simple yes or no; the former allow much freer expression of opinions (for example, what types of food do you eat for breakfast?). Many questionnaires also include attitude-scaling questions which seek to identify degrees of satisfaction with products/services: 'Strongly agree, Agree, Disagree, Strongly disagree'; 'rank these characteristics in order of preference' (see, for example, Box 3.1).

Care must also be exercised in the administration of questionnaires, to ensure that results are not distorted by prompting, misunderstanding, or ambiguity. In the past, analysis of the results was difficult because of the time involved and interpretation difficulties, but this has been substantially simplified by computer software. Direct input of responses into the program is even possible instantaneously over the telephone, or into a lap-top computer as the interview is conducted.

SEGMENTING MARKETS

Segmenting markets on the basis of market research data is not difficult, but the results must be accurate if they are to provide an accurate product specification and a reliable basis for the financial and operational feasibility appraisal. The ways in which customers are distinguished from each other is important, and the criteria used for segmentation must be appropriate: only those which relate to purchasing behaviour and can be put into action will be useful to the manager.

There are four aspects to the segmentation process: segment identification, measurability, accessibility and appropriateness. The process is iterative, and it may have to be repeated several times if the segment initially identified does not fulfil these requirements. For example, if it proves impossible to communicate with the segment or it is too small for financial viability, the identification stage will have to be repeated, perhaps broadening the definition of the target group to increase the size and permit a new form of communication.

SEGMENT IDENTIFICATION

It is normal practice to begin with the total market and progressively narrow the field to arrive at a targetable segment. For a potential exporter this may mean starting with a country or a group of countries (for example, the EU, Australasia, South America). Within a country it means starting with published statistics about a national or regional demand, and ending up with a small customer group who share a set of characteristics from which their product requirements may be inferred.

A segment may often be shown to exist and the size of it may be known in gross terms: for example, the number of EU consumers who express interest in organic food is known, and represents a potential market of several million. This is unhelpful unless more information can be obtained about other customer characteristics (who they are, where they are, what their requirements are, and whether they will actually buy). The data may be difficult to collect, particularly if the product is likely to appeal to only a very small niche market.

In situations where well-defined segments have been identified and information is more readily available (vegetarians, high-income gourmets, single-parent households on tight budgets) the market may already be over-populated with other suppliers. The segment may therefore have to be disregarded, or a small gap sought out which a differentiated product could supply. A third strategy is to try to split the market into smaller pieces. Fruit canners, for example, were able to split their market into a 'canned in syrup' and a 'canned in juice' segment on the strength of customer interest in healthy eating: research identified a change in demand which generated a new segment. The subsequent introduction of 'canned in light syrup' indicated even further segmentation of the market which acknowledged an intermediate category of consumer not satisfied by the other two offers.

Continuous monitoring is therefore necessary to identify new segments which are emerging or which earlier research did not pick up. Wide product areas must also be covered to detect demand shifts between substitute products and variations in product presentation. For a fruit canner, for instance, a consumption shift from canned to fresh fruit is obviously important, but so is a shift from fruit to breakfast cereals, dairy desserts, ice-cream or even cheese. As ever, opportunities as well as threats need to be identified: in the last example, a move towards desserts may generate a new market segment for the fruit canner for the supply of fruit chips to dessert manufacturers. Within segments already supplied monitoring is necessary to identify demand changes which, if unobserved, could lead to declining sales unless the product is not completely re-designed and re-launched.

SEGMENT MEASURABILITY

The size of an identified segment is critical, both in actual terms and in relation to the growth expectations of the business. The segment may be too small to be economic, or too large to be supplied without over-stretching resources, leading to the possible collapse of the business. For example, farmhouse producers often wish to target the customers of a particular retail chain, but if sufficient supplies are required to stock the entire chain this will entail an expansion of production and increased risk. Without a long-term contract it would therefore be unwise to sell to this segment.

The total market size for most product groups and for fairly large, well-defined geographic and demographic segments is easily discovered from published sources. Research companies sell data derived from consumer surveys and supermarket scanning returns which may be sufficient to identify a small segment, to be served with a branded niche product. The size of most market segments can therefore be calculated fairly easily. For example, using published demographic figures, a potential producer of a luxury vegetarian baby food could estimate the total likely market for all vegetarian baby foods in a given geographical area:

Total number of households in area	670,000
Households with children under 1 year (4.5% total population, so 4.5% of 670,000)	30,150
Households in socio-economic groups A & B (19% total population, so 19% of 30,150)	5,728
Vegetarian households (6% of 5,728)	344
Annual value of market segment (£520 x 344)	£178,880

Primary consumer research would then be required to establish the producer's potential share of this market – say 10 per cent, giving a market segment to the value of £17,888. If this were considered too small to be viable, a marketing strategy would be designed to achieve a market share of more than 10 per cent, or the geographical area could be extended.

SEGMENT ACCESSIBILITY

The feasibility of effective communication with the segment must be established as a condition of segment viability (since there is no point in targeting it if it does not know the product is available or cannot gain access to it). The right distribution channel to make the product available to the right customer must therefore be determined, and the producer's ability to access it. The feasibility and cost of effective promotional communication with the customer group, via the right media, must also be investigated. If the product is targeted at a small group, the need to aim the rifle shot accurately is particularly critical.

Communication to end-users and intermediary buyers is achieved through such media as newspapers, TV, radio, trade press and the Internet, and is backed up by sales personnel. In most countries there is such a wide range of general and specialist publications aimed at consumers and trade buyers that most market segments are reasonably easy to access. For example, there are vast numbers of cookery and healthy eating magazines, and trade press covering all food commodity, catering and retailer groups. Media providers have information about their own customers which they will provide to clients wishing to advertise. However large or small the segment, it should therefore be possible in most countries to identify the right medium, though the help of a communication agency may be necessary in choosing between near-substitutes.

SEGMENT APPROPRIATENESS

The market segment chosen must be appropriate to the business. The obvious criterion already noted is scale, but there are other considerations. Small producers are generally wiser to restrict initial sales to their local area and extend their geographical reach only when this has been satisfied. This derives the maximum benefit from any local reputation while reducing distribution costs. All the marketing activities must also be appropriate to the targeted segment. High-quality, premium-priced luxury products need a suitable segment, and if no such segment exists, the only option in the short term is to change the product to one for which a viably-sized segment exists. From this base it may then be possible to create an up-market segment for a higher-quality product.

SEGMENTATION VARIABLES

The segmentation variables by which a customer group is defined will normally suggest themselves in the course of market research. They will include readily identifiable factors like income, location, demographics (age, family

Demographic Variables

Sex
Age
Marital status
Family size and background
Race/ethnic group
Education
Occupation
Income
Religion
Home Ownership
Socio-economic class

Geographic Variables

Region
Urban/suburban/rural
Population density
City or county size
Market density
Residential location
Housing type
Climate
Terrain

CONSUMER MARKETS

Benefit Variables

Usage rate and volume
Product benefits
Consumer need satisfied
Technical aspects
Price sensitivity
Brand loyalty
End-use
Benefit expectations

Psychographic Variables

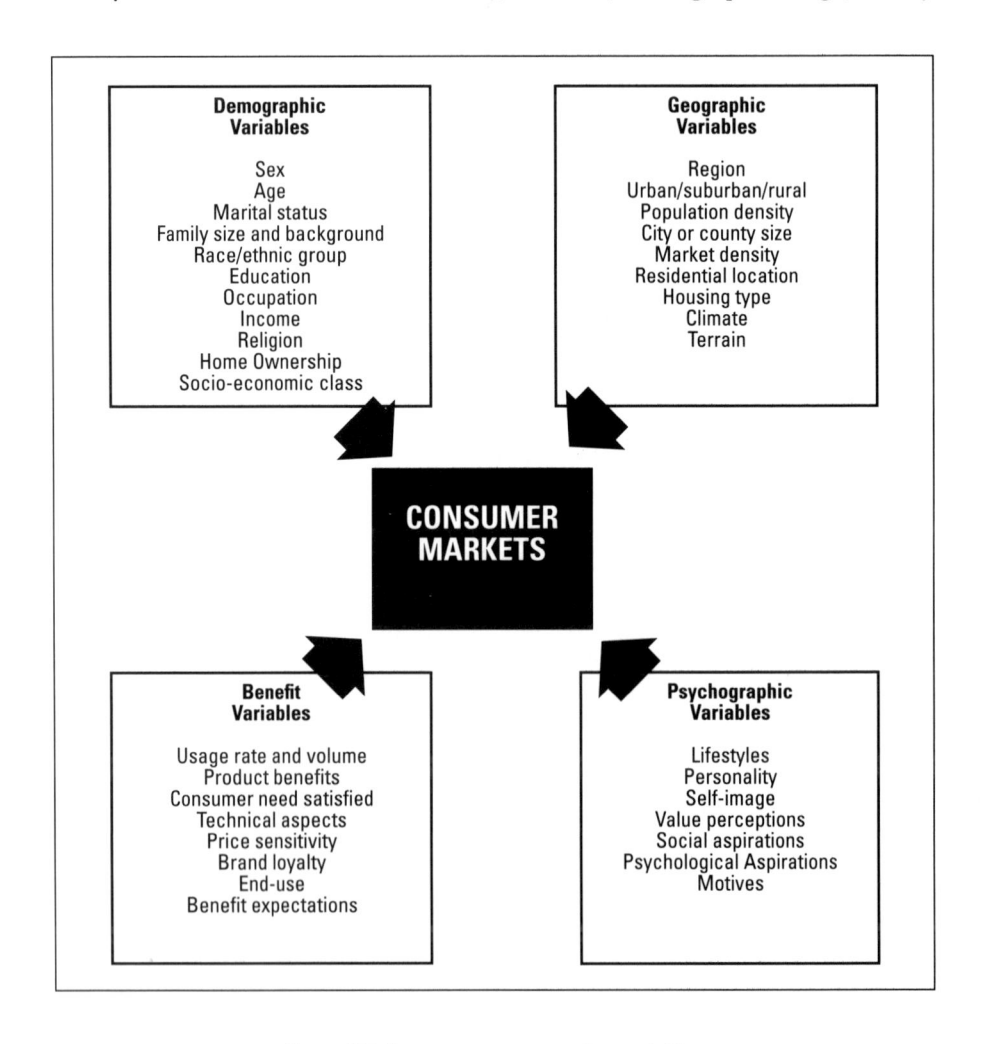

Lifestyles
Personality
Self-image
Value perceptions
Social aspirations
Psychological Aspirations
Motives

Figure 6.7 Consumer segmentation variables

size etc.), and less readily identifiable psychographic factors like status consciousness, brand loyalty, attitudes to convenience, eco-products, healthy eating etc. (Figure 6.7). Though the latter are more difficult to identify, the effort must be made because these are the variables which often determine a customer's choice of one product over another.

Volume segmentation

Volume segmentation seeks to identify potential customer segments in terms of the amount of product used (light, medium or heavy). If this can be related to another customer characteristic about which other information already exists, narrow targeting of this group can be achieved via promotional offers, price incentives etc. In one survey, for instance, heavy users accounted for over a third of baked bean consumption. If students could be identified as heavy users, they could be targeted easily at their place of residence or local stores (a known geographic variable), with promotional offers likely to appeal to students and bulk users (money-off next purchase). If heavy users of agrochemicals are arable farmers with larger than average farms, communication targeted through *Big Farm Weekly* is probably better than through *Farmers Weekly*.

Geographic segmentation

Food consumption and other data bases show wide variations in regional demand. The reasons for these variations are complex, but they include factors like climate and its effect on product availability, population composition (ethnic and cultural differences, age structure, income distribution, urban/rural community etc.), plus local tastes and habits. Even in a small and fairly homogeneous country there are large regional variations in the type of food consumed: in the UK, for instance, regional variations of plus or minus 30 per cent of the mean consumption of a given product are regularly and consistently found.

Geographical location is therefore a sensible and a relatively easy basis for segmentation, using categories such as urban/rural, inner-city, a 25-mile radius of the capital city, as well as named places. Figures are readily available (government statistics), so identification and measurement are straightforward (for instance, higher or lower income areas within a town). Accessibility is also relatively easy, since most areas have one or more local newspapers, radio and TV channels. (Commercial radio and TV companies are incidentally a major source of market information for products which are not as well documented as food.)

Regional segmentation is often related to distribution. It may be necessary to limit distribution to certain areas for logistic/cost reasons, or a retailer may be used who only operates in certain regions. In very large countries like the USA and Australia such a policy may be essential to run a cost-effective system. In small countries regional segmentation is ideally suited to the small business, and may be the only way to operate.

Demographic segmentation

Demographic factors are probably the most widely used segmentation variable, yielding a range of population data like socio-economic or age group, family size, life cycle stage (retired or young households). Food consumption statistics usually incorporate much data of a demographic nature, as does census information. These characteristics may be combined to define a segment in terms of family size, income level, age range: for example, two working adults aged 30 and 29, with two children aged four and two years, income £50,000. This would give a fairly precise identification useful, say, to a business wishing to sell high-priced foods for young children.

The problem then is communication, since such families will be regionally scattered, though possibly more densely located in one region rather than another. The combination of demographic with geographic data, together with other information about social class and cultural background, is thus desirable. The task is made easier by the existence in many countries of standard geo-demographic classification systems based on census data, readership, TV and other consumer surveys. In the UK, for example, ACORN (A Classification of Residential Neighbourhoods) relates population characteristics recorded in the national Census (including age, sex, socio-economic group and occupation) to the Census enumeration districts from which the information is drawn. Consumers can therefore be grouped by socio-economic group and residential area: for example, wealthy young male achievers living in suburban areas; affluent grey couples in rural areas; skilled workers in home-owning areas; older singles in less prosperous areas. These groups are further broken down, and are used by market researchers as a basis for detailed primary research.

Psychographic segmentation

Psychographic segmentation identifies the psychological benefits sought by customers, and their attitudes and motivations in buying products. It seeks to classify customers by their values, opinions and personal characteristics. For example, one survey of 20,000 consumers which categorised respondents by healthy eating attitudes generated seven segments whose size was estimated (Table 6.1.). By incorporating other customer variables into this analysis (existing purchasing patterns, shopping outlets, animal welfare concern etc.), different product information could be communicated via different media.

Progress in identifying targetable psychographic factors has been limited by the difficulties involved, although much experimental work has been done. Socio-economic classifications like ACORN, and other commercially published lifestyle classifications, may suggest characteristics which can be investigated. These are generally constructed on the basis of information collected by national census investigation, updated on a regular basis and supplemented by private research. Primary research must always supplement these sources, however, because socio-economic and other variables are not necessarily a reliable indicator of purchasing behaviour. For example, consumers resident in up-market areas do not necessarily spend more on food, or choose branded products.

Brand loyalty

Many consumers exhibit strong loyalty to manufacturer and retailer brands, which extends across product groups (for example, food and non-food products). This provides a potentially useful segmenting variable, since these customers may in principle be expected to share other characteristics which might be helpful in identifying product attributes and communication strategies. The necessary data may be collected from an existing customer base: for instance,

> **Survey identifies seven types of food faddist**
>
> *Superfits (13%)*
> Middle class, living in SE and SW England; a significant proportion of this 'well-informed and enthusiastic' bunch are working class.
>
> *Younger concerned (14%)*
> Less determined, more confused, because committed to healthy living, but insufficient knowledge to make rational choices, and their chaotic lifestyle undermines their good intentions. Suspicious husbands obstruct wives' good intentions; mothers are prepared to compromise for their children.
>
> *Older concerned (14%)*
> More determined over 45s, avoiding high-risk foods through fear of illness, obesity or premature death.
>
> *Selectivists (17%)*
> Young-to-middle aged, middle class, resident in the south of England outside London. Interested in family care and looking good, have changed eating habits as manufacturers provided healthy alternatives.
>
> *Traditional healthers (12%)*
> Slightly older and disapprove of fashions. Believe in balanced diet and prefer natural products.
>
> *Dismissers (13%)*
> Reject the healthy living arguments and consume fast food, take-aways, cakes and fizzy drinks.
>
> *The untouched (17%)*
> Remain 'supremely apathetic'; nearly half are over 65 years of age.
>
> Adapted from a survey of 11,000 housewives by Advertising Agency D'Arcy Benton & Bowles

Table 6.1

via loyalty cards, in-store customer surveys, or the short questionnaires which accompany many durable goods purchases (for example, guarantee/warranty cards with electrical appliances). This has long been standard practice for manufacturers and large companies, and it is a potentially valuable segmentation approach for a small company with its own established customer base and reputation (brand).

CONCLUSION

Demand is typically heterogeneous, and a business which can identify a targetable niche may gain a competitive advantage. The difficulty for farmers in responding to this is that they generally experience a *derived demand*: consumer demand interpreted by intermediaries. Where there is accurate and prompt feedback of consumer requirements this is no problem, but some channels and intermediaries are poor communicators and may be slow to identify new product opportunities. A knowledge of consumer as well as intermediary markets is therefore necessary both to identify segmentation opportunities and to select the right marketing channel/intermediary to achieve them.

CHAPTER 7

The product decision

A product is a good, a service or an idea, or any combination of the three which delivers a defined customer need. The product decision identifies exactly which product attributes must be provided in order to target an identified demand, and the management implications of supplying it. Since demand is dynamic, and competitor offers and the marketing environment also change over time, this is not a once-and-for-all task: product development is a continuous process which responds to (and ideally anticipates) changes in demand and competitor activity.

In many farm businesses the basic product decision – what to produce – is limited by the natural environment and physical situation. This is obvious in the case of agricultural enterprises (hill farms cannot grow crops), but it may also be true of non-agricultural ventures which would be viable in other situations. Remoteness from population centres may undermine the financial viability of a leisure enterprise, for instance, or it may add cost and cause logistical problems for a farm-based food processing enterprise (distance, inadequate access for large distributors' vehicles). The product choice in many farm and rural businesses is further restricted by:

- legal, social and land-use constraints which either prevent certain types of development altogether, or increase the costs and operational problems
- designated area planning constraints which may entail higher building costs, impose access restraints, limited opening times, quiet activities, etc.
- compliance with public health and hygiene requirements which may entail prohibitive plant, equipment and labour costs that rule out a processing enterprise.

However, it is rarely the case that existing products cannot be modified and re-positioned in the market to add value and gain a competitive advantage. New products may also be introduced to reduce dependence on a narrow farming base.

This chapter concentrates on ways of maximising the revenue from existing products, and outlines the process of new product development. Product range is not discussed here, since it was considered in Chapter 3, but readers are reminded that individual product decisions must always be made in relation to the consistency and credibility of the total business offer, their potential impact on overall revenue, costs, and profitability, and their implications for resource allocation.

PRODUCT DESIGN

Chapter 6 explained that a customer purchasing any product buys a combination of functional and psychological benefits which can be modified to encourage purchase. The product supplied by a milk producer-retailer might therefore include:

- *functional attributes*: fat content, organic source, different flavours
- *functional services*: doorstep delivery, monthly accounts, eco-packaging
- *psychological benefits*: reliability, friendliness of deliveryman, product assurance.

Functional attributes and services represent the 'core product': the minimum necessary for customer satisfaction (Figure 7.1). A doorstep milk service must include milk and doorstep delivery; farmhouse B&B must include sleeping and washing facilities plus basic warmth and shelter. To that extent, all doorstep suppliers and all farmhouse B&Bs are identical. Reliability and personal attention, en-suite facilities and a higher standard and style of furnishing add psychological benefits which differentiate and 'augment' the product, and allow a premium to be charged for superior provision.

For most food products, firms at different points in the marketing channel supply different attributes. Primary producers and manufacturers tend to focus on intrinsic product characteristics while distributors are concerned with add-

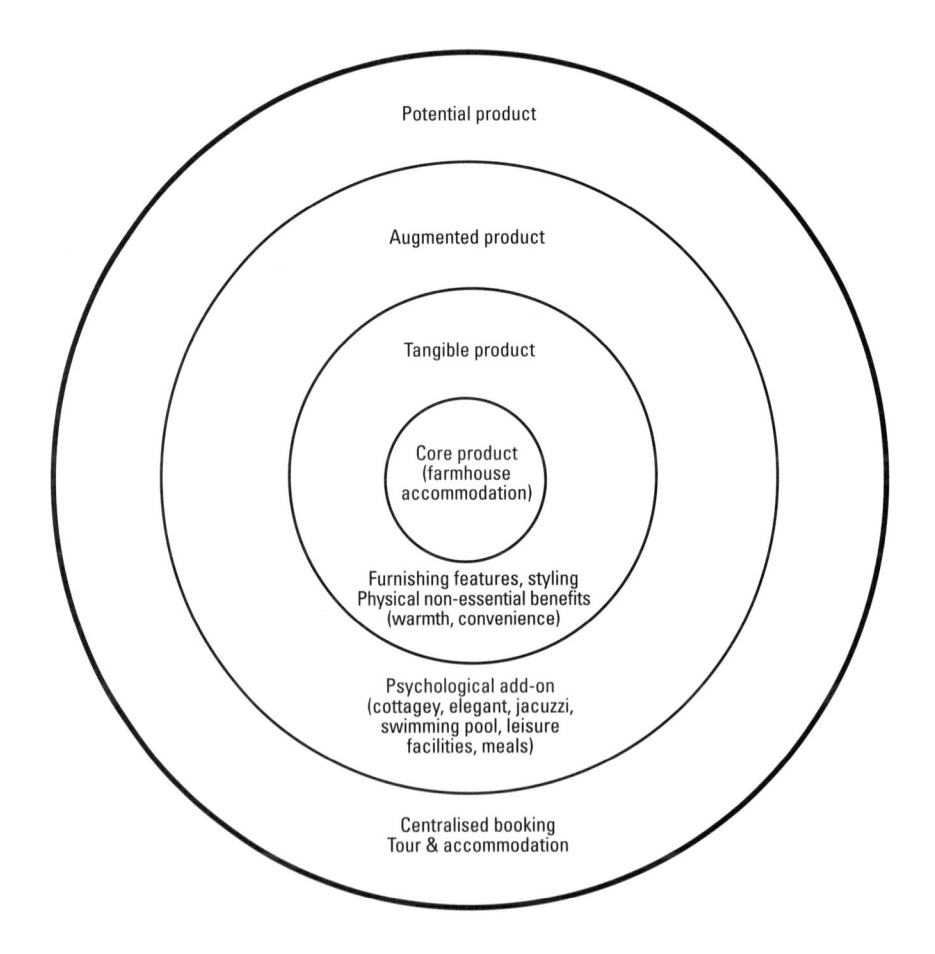

Figure 7.1 The elements of a product: e.g. farmhouse accommodation

on services. Everyone in the channel is concerned with price determination and promotion, and the producer-retailer is responsible for all four. Farmers are typically concerned with what attributes they must and can provide, and which downstream intermediaries will best supply the rest, so that they benefit from the combined marketing effort.

PHYSICAL AND SERVICE ATTRIBUTES

Decisions about the physical aspects of a product include:

- the initial choice of product to supply an identified demand: for example, the right crop variety or livestock breed; serviced or self-catering accommodation
- the production implications: fertiliser/feeding regime, terminal sire; daily/weekly cleaning schedule, prompt maintenance, regular refurbishing of accommodation.

Most farmers have no difficulty in identifying with this aspect of marketing management, in farming or alternative enterprises. Where they often err is in giving insufficient attention to all the functional aspects of a product or service because their product specification meets their own expectations, but not that of their customers. Many commodity producers are dismissive of 'pernickety' buyer specifications which reflect a level of demand they fail to appreciate, which provides opportunities to add value to a core product. In alternative enterprises like farmhouse accommodation and leisure/recreation facilities, the same attitude may result in a product specification which is too narrow. For example, self-catering units should generally have the latest microwaves, automatic washer-dryer and dishwasher, and be serviced to a high standard; leisure facilities must have super-clean, modern toilets and car parking facilities, professional-looking signs, and courteous staff.

Market research will frequently establish that the service aspects of a product are as important to the customer as its physical attributes. Many tractors, for example, are bought on the strength of spare part availability and after-sales service as much as for their physical characteristics and performance. Similar guarantees of product attributes and performance and after-sales service must be regarded as an integral part of any product mix: for example, replacement and/or reimbursement policy on faulty products; advice on product use; upgrading and updating facilities; efficiency and courtesy in dealing with customers.

PSYCHOLOGICAL ATTRIBUTES

An excellent physical product plus all the right service characteristics is rarely sufficient to guarantee customer satisfaction and competitiveness. Psychological attributes are frequently the factor which determines customer choice between near-identical or substitute products and allows a premium to be charged for perceived added value: for instance, beef with a product assurance; potatoes, graded and packed by size for baking, or in a processed form ('luxury mashed potatoes').

Psychological benefits frequently explain why customers will pay more for the same product in different outlets or more expensive packaging. They are also the principal means of overcoming habitual buying behaviour, by catching attention through an innovative product offer or imaginative promotion. Once a customer is acquired, reinforcing action to establish a new pattern of habitual purchase then relies heavily on psychological benefits delivered through product assurances, price-led sales promotions and regular product enhancement. The psychological content of a product thus has significant marketing potential.

This is true of essential as it is of non-essential consumer purchases, and it is true whether the customer is a consumer or an industrial buyer or distributor. Intermediary buyers considering competing offers look for the psychological value to the consumer in addition to intrinsic product attributes (though they are unlikely to use these terms). Professional buyers themselves are also not immune to the psychological content of a product. Many retail buyers, for example, have a poor opinion of some agricultural marketing channels which influences their perception of the product. Multiple retailer meat buyers, for instance, have always associated livestock marts with unreliable quantity and quality, and animal welfare fears which they know to be important to their customers, and this motivated their search for direct sales and producer alliances.

Where personal (face to face) selling is involved – to professional buyers as well as consumers, the psychological content of a product includes the manner of its delivery. At industrial buyer or distributor level, the professionalism of the encounter and the ease with which the buyer can deal with a supplier is critical in gaining product access and sales. At consumer level (and particularly so for service products), courtesy, efficiency, and ease of transaction are equally critical. The attitude and manner of sales personnel contribute substantially to a satisfactory sale, repeat purchase and personal recommendation. In the case of services, everyone involved in providing the service is a salesperson: for example, in a farm accommodation or catering enterprise, this includes cleaner, waitress, farm staff and family members (including any who are not directly involved in the enterprise). Where retail premises or vehicle distribution are involved, their quality and 'style' express in a tangible form the intangible psychological benefits of the product offer, and thereby make a strong contribution to customer satisfaction.

The psychological content of a product is summed up in the word 'image', and for some products in consumer markets, image is all. The classic example is *Coca Cola*, which represents a lifestyle image that consumers buy along with the contents of the can. In this case the psychological benefits are at least as important as the physical product, and the same is true of many food products which supply psychological benefit in addition to nutrition.

The obvious example is organic food, which cannot objectively be shown to be superior to other products, and most consumers cannot distinguish in blind tastes, but some consumers believe to be superior. The same is true of products marketed as free-range, natural, environmentally-friendly, humane, fresh from the farmhouse kitchen; all these add value to a routine purchase by *augmenting* the standard offer. A major objective of many generic marketing campaigns conducted by farmers' organisations is to add value in this way to an undervalued commodity. Meat advertising campaigns have been particularly effective in counteracting consumer health fears and adverse publicity by a combination of quality assurance and psychological appeals to family values, a loving relationship, a healthy masculine image etc. (Plate 10).

The same principle applies equally to alternative farm enterprises. A farm accommodation enterprise can appeal to psychological benefits like peace and quiet, idyllic surroundings, freedom from urban fears. The aesthetic appeal of furnishings and the wholesomeness of home cooking contribute to this, and may allow a premium to be charged for the right product. The novelty, fashionability or status of a leisure pursuit can be stressed to promote sales. Every effort should therefore be made to convey these intangible benefits via well-designed promotional material (critical where prior inspection of the product is impossible), and to ensure that service delivery is consistent with the promise (no muddy farmyard or noisy farm operations to disturb the rural idyll).

PRODUCT BRANDING

The ultimate expression of a product with a strong psychological content is the brand: a coded message which communicates the benefits customers can expect to derive from purchasing a product. It identifies known product attributes with a known supplier, and provides a product assurance which guarantees that the product delivered will be consistently of the same standard. Only small deviations from the standard are likely to be tolerated, and only then on the understanding that any such change will be reliably managed without loss of customer satisfaction (for example, the substitution of inputs in the case of shortage).

Branding reduces the customer's effort and the risk involved in buying non-branded products; for the supplier it creates a customer base loyal to a brand, which reduces the marketing effort necessary to generate sales. The power of a brand is readily communicated by names like Marks & Spencer, Kellogg's or John Deere, but branding is not the prerogative of large companies with large advertising budgets. A brand is simply a reputation for supplying a consistently reliable product, which many farmers establish over the years with particular buyers. The problem for raw material suppliers is that the brand does not extend beyond the firsthand buyer, and this exposes them to the risk of input substitution at the whim or the requirement of their firsthand buyer. This was demonstrated by the BSE beef crisis of 1996, when producers all over Europe could find no market for an unbranded beef product, and even branded products had to struggle to restore customer confidence by actively supporting their brand.

In the absence of producer brands for agricultural raw materials, commodity and producer organisations carry out generic promotion which builds a reputation for an entire sector's output, from which individual producers benefit. The effectiveness of generic promotion is limited, however, if the brand is not carried through to retail level, and there are no means of ensuring that what is promised is actually delivered. A 1980s campaign to promote English Cox apples, for example, was very effective in establishing the brand and stimulating consumer demand for the home-grown product. However, individual producers who marketed variable produce with varying levels of service and efficiency did not deliver a uniform product, or match the competition from better graded and better presented imports. In the 1990s a similar promotion of Bramley apples was firmly underpinned by product quality control and marketing discipline to make the brand credible, and the result was measurable within a season in increased consumption and strong demand (Plate 11 and Box 7.1).

In some farming sectors, disciplined production and marketing have always been a normal requirement for anyone who seeks a supply contract. The obvious example is milk, where quality-linked price incentives and penalties (and termination of contract in the last resort) are routinely used to achieve the standard necessary to underpin a brand like Milk Marque or Unigate. The need for similar assurance is now recognised in the growth of producer marketing groups and individual producer-buyer relationships which work to a standard product specification, with premia for quality achieved and penalties for default. This is an absolute pre-requisite for a branding strategy, which seeks to build customer recognition and repeat purchase by a combination of distinctive product packaging, corporate imaging, and in-store and media advertising (Plates 12-14). For the individual producer-processor, retailer own-labels provide an alternative way of enjoying the benefits of a brand without the establishment costs, by producing to specification.

Bramley records possible

Apple growers have claimed that if multiples resist the temptation of price promotion during the forth-coming Bramley Week campaign, it could be the most profitable retailers will experience since the event began five years ago.

Bramley Group's Ian Mitchell says, 'This year retail prices for Class 1 fruit have been around 59p/lb and demand has been buoyant. We are confident we can stimulate demand without the need for price promotions to move volumes, particularly in a shorter crop year.'

The theme is 'Back to the future', backed by recipes in consumer media. Additional activities are linked to Comic Relief, and Bramley races where 22,000 junior schools have been asked to participate.

The Grocer, 25 January 1997

Box 7.1

Brand consistency

Effective branding depends, then, not just on setting a reliable product standard in the first place, but on strict adherence to the standard set (brand consistency). This should not be difficult for a small owner-managed business to achieve, but it may become difficult as the business grows and responsibility is delegated to employees. The problem is often encountered, for instance, when farmhouse processing is scaled up or transferred to new premises which are less directly under the owner's personal control.

Once a brand is established, everything marketed under that label must also live up to the guarantee it represents. A single product which failed to live up to a firm's reputation for known quality and reliability could damage customer confidence and sales across its entire product offer. For example, a marketing group which supplies quality *finished* sheep risks its reputation by marketing poor *store* stock. A farm shop which sells quality vegetables must supply equivalent quality throughout its product range and service. Quality farmhouse accommodation must not be compromised by inferior or poor-value meals. The same warning applies where a business diversifies into new enterprises, and seeks to transfer the benefit of its established reputation to unrelated products or enterprises. For instance, a farmhouse processor of ice-cream who opens an ice-cream parlour or farm restaurant must deliver the new product (a service, not ice-cream) to a standard which will not damage the existing brand.

THE PRODUCT LIFE CYCLE

A major objective of branding is to build customer loyalty to a supplier as well as a product, because demand for a product inevitably declines over time. However good a product is, a time will come when it is outclassed because:

- superior products of the same type enter the market (better-presented meat products; packed, prepared vegetables; en-suite farmhouse accommodation)
- a whole class of products is superseded by a new one which performs the same function better (the binder is superseded by the combine; frozen by chilled foods)
- customer demand changes (substitution of spreads for butter, skimmed for full-cream milk; emphasis on convenience, green products)
- totally new products are introduced to satisfy new demands (self-peeling fruit; up-market holiday villages combining leisure and accommodation).

Consequently, all products have a limited life cycle which reflects the dynamic nature of the market. The impact of this varies by business sector and by commodity: some products have a short cycle (particularly fashion products – including many food products) while others are still in their growth phase years after their introduction. All products nevertheless show the same pattern of growth and decline (Figure 7.2). Most of the factors which influence this are beyond management control, but an understanding of the product life cycle, combined with good marketing intelligence and internal sales and revenue data, will indicate when and where action is needed to modify or abandon declining products – or (for strongly branded products) to reinforce customer loyalty through vigorous promotion and re-positioning.

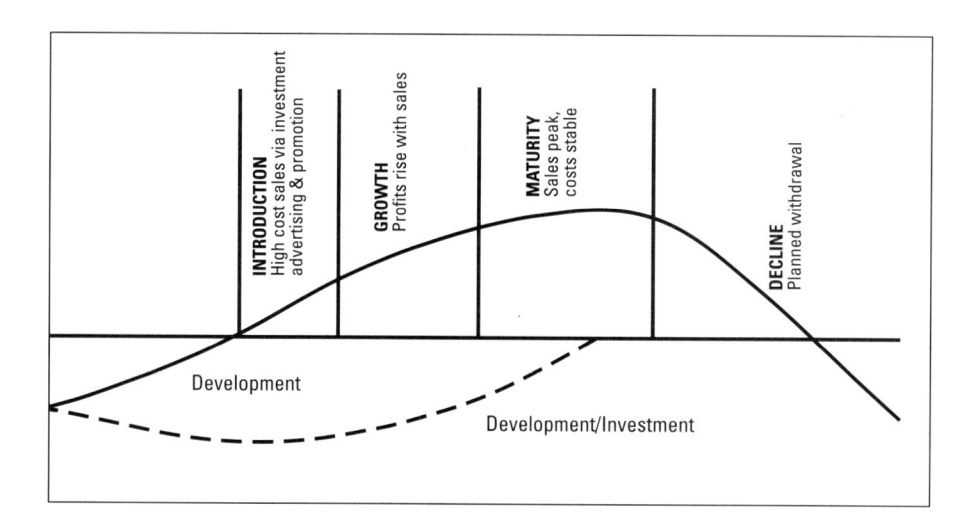

Figure 7.2 The product life cycle

MANAGING THE PRODUCT LIFE CYCLE

An awareness of the stage at which various products are in their life cycle is essential to the management of the marketing mix. It is also relevant to the product portfolio decision (Chapter 3), in identifying where resources and effort need to be targeted and where they would be optimally deployed – not necessarily the same.

In the *introductory* stage of a product's life, market development is all-important, so priority must be given to:

- the establishment of a distribution channel
- active promotion
- (possibly) competitive pricing
- customer monitoring to ensure that necessary product modification is promptly implemented.

At the *growth* stage:

- increasing sales may strain productive capacity and financial resources beyond the planned capacity
- competitors' reactions must be monitored so that product modifications necessary to match any competitive offer are rapidly identified.

At the *mature* stage the object is to prolong this phase of the product life cycle to maximise the return on the initial investment. This means:

- defending the market position by further product improvement, segmentation, price incentives, other promotions, and product line extensions (modifications of existing products: for example, new yogurt or ice-cream flavours; en-suite accommodation)
- transferring management effort and resources into other products at an earlier stage of the product life cycle, and/or introducing new products in anticipation of declining sales.

In the *terminal* stage when products start to decline, the basic choice is:

- whether it is possible to revive them by a major re-design and re-launch
- whether this represents a misdirection of management effort and investment which could be better deployed elsewhere in the product portfolio.

As Chapter 3 explained, product deletion may have adverse effects on other products if it throws doubt on the total offer or business credibility. The decision to abandon a product therefore needs careful management, the principal considerations being: to phase it out without adverse impact on the overall marketing strategy, to exploit any remaining strengths, and to avoid bad PR (unfavourable perceptions). This is easier if alternative products are already on stream, and the phasing out of one product is off-set by the growth phase of others.

Financial implications

The product life cycle is also helpful in clarifying the financial implications of product management. In the product development stage sales tend to be low while investment and management costs are high. As the product is accepted and sales grow rapidly, profits rise as development costs are recovered, and unit costs decrease as production expands. In time sales level off together with profits, and both will normally decline. The implications are that:

- financial planning must be related to a realistic sales and costs forecast over the whole product life cycle
- product modification and/or substitution must be planned well in advance to even out revenue
- investment in new products should start before profits begin to fall.

The process of product development must therefore be continuous, and the object in a multi-product business is to achieve a portfolio of products with over-lapping life cycles and complementary investment and management patterns (Figure 7.3). In very rare cases it may be possible to do nothing, in the hope of surviving longer than everyone else and thus preserving a small market niche, usually in a small specialist market. Although this *can* become profitable it is a rare achievement, and in most situations the ability to modify and relaunch products to meet changing technological and market trends is critical to business profitability and sustained viability.

The appropriate response will vary in different industries and for different products in the same industry, but two broad strategies exist: the maximisation of revenue from existing products, and the introduction of new revenue-generating products/enterprises.

The assessment of these alternative strategies will depend primarily on financial criteria (increased revenue or reduced costs), but the assessment must

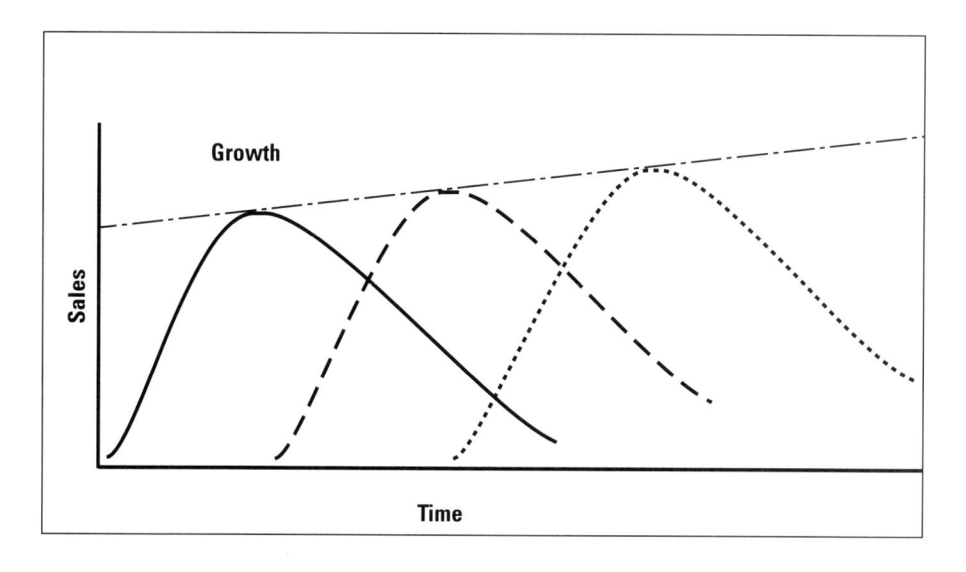

Figure 7.3 Overlapping product cycles sustain business growth

also reflect other business objectives: for example, the need for additional enterprises to employ family members or, at the other extreme, to reduce the workload while sustaining the revenue-generating capacity of the business. It is also important to distinguish between short-term and long-term effects. If short-term benefits are sought it is rarely sensible to make more than marginal changes to the existing product range because the transition costs of a major business redirection are likely to exceed short-term benefits achieved. For long-term growth, it may be necessary to accept the added short-term costs in anticipation of much greater long-run profits.

MAXIMISING REVENUE FROM EXISTING PRODUCTS

In most businesses the priority is to improve the revenue-earning capacity of current products by better *positioning*, achieved through product and marketing modifications which match customer demand better and improve customer perceptions relative to competing offers. The existence of buyer specifications greatly simplifies this task, since producers simply have to deliver to a given specification. In all other situations marketing research is necessary to provide information about existing and potential customers and their product and service requirements, competitor offers, and more general market data which together provide a comparable delivery specification. In the light of this and the internal business appraisal it is then possible to identify feasible and cost-effective action to:

- find a new market for an existing product
- modify the existing product by changing existing attributes and/or adding new ones
- improve customer perceptions of the product.

These are complementary, not alternative, strategies, and action on all three fronts will typically be required to improve the return from a product. A new market for an unmodified product may be identified, but in the process of supplying it opportunities for product modifications may be recognised which would enhance revenue or sales. Production changes also invariably need supporting by promotion, to ensure that customers perceive the improvements.

A NEW MARKET FOR AN EXISTING PRODUCT

The first strategy most managers will consider is to research the market for new customers for an existing product. A potato producer may be able to improve revenue simply by targeting separate markets which exist for different sizes, different seasons and different qualities, without any product modifications. The simple grading out of baking potatoes is a well-known example where the retail price was multiplied many times over. Seasonal gaps in the market traditionally presented an opportunity which worldwide sourcing has partially eroded, but there are still market segments for home-produced 'new season' products, often at a high premium, reflecting the scarcity value and production cost.

Supply patterns and production schedules may also be modifiable to *extend* the season: for example, through the introduction of new varieties and/or management systems. Processing markets (potatoes for crisping, vegetables for freezing, apples for juicing) can contribute to improved revenue by allowing high-quality output to be targeted at a premium market while finding markets for lower quality outgrades which would otherwise be unsaleable, or only at a substantial discount.

Kentish Garden seeks to shift strawberry season

English strawberry growers are being encouraged to shift their seasonal pattern by Kentish Garden, the UK's largest soft fruit cooperative. Because it has had a profitable trading year, with sales in excess of £20 million, the organisation was able to offer its 64 members a three-year deferred payment scheme worth £750,000 to grow more fruit under plastic tunnels. This costs around £10,000 an acre.

'As well as producing more Class 1 maincrop strawberries earlier when they were wanted in June, the portable structures can be used later in the year to protect later-maturing Elsanta strawberries,' explained the Managing Director. He said the initiative had been welcomed by his supermarket customers, and would result in greater confidence in autumn-grown fruit.

The Grocer, 21 September 1996

Box 7.2

Examples of other likely opportunities which need investigation are:

- geographical gaps: relatively easy to discover in domestic markets, or in export markets, but may need joint producer ventures
- demographic gaps: for example, a singles market for farmhouse accommodation; one-person packs of mixed fruit and vegetables for single customers
- income groups not served by existing products in the marketplace: for example, pensioners on limited incomes who would like a prestige or high-quality product in an affordable size.

All these call for marketing rather than production skills – which may explain why the return from the product is unsatisfactory in the first place. For a manager whose strength lies in production, and who has limited time for market research, it may therefore be wiser to look first for production modifications which will supply existing markets better.

PRODUCT MODIFICATION

Product modification is a major competitive force in manufacturing industry and in most service sectors, and the driving force of most marketing programmes. Manufacturers introduce more sophisticated products; retailers offer a wider range of products and better service: longer opening hours, bigger car parks, home delivery. Farmers can offer improved varieties of basic raw materials or higher health and hygiene standards, better quality products, increased levels of service: sorting, grading, part packing etc. The psychological benefit achieved by better packaging design often allows a small premium to be charged: for example, mineral water in coloured glass bottles with a designer label in place of plastic bottles (Plate 20). Relatively minor changes of this kind have changed the whole market for some products and justified the additional costs incurred, by moving them out of a crowded, highly-competitive, price-led market into an up-market sector.

The greatest scope for product change lies in processing agricultural raw materials into consumer food products, which may also employ under-utilised resources (human or other). The added costs and management commitment are greatly under-estimated, however, and this is rarely the profitable option it is thought to be. Before venturing down this road a wise manager will therefore consider many less radical changes to products which will improve revenue or secure a market niche, without entailing the same financial and personal commitment. For example, different livestock breeds and crop varieties with different quality attributes will underpin a market segmentation strategy. Cereals can be grown which were developed for specific manufacturing purposes (bread-making wheats, malting barley). Different varieties of potato may be planted to supply crisp manufacture or frozen chips; others have been bred or re-introduced to meet particular consumer needs (better taste, salad use). The right choice of breed or variety is therefore the first step, combined with a husbandry regime which will ensure that the output meets the required intrinsic standard.

Sometimes no change in a product's intrinsic attributes is necessary, but a production change will improve its psychological content and customer image. The obvious example is conversion to organic farming, or, less drastically, the introduction of more extensive and environmentally-friendly regimes, or a high level of animal welfare. The value of an existing product can also be improved by stressing a wider range of uses to which it can be put. The use of potatoes in salads and as a main ingredient in vegetarian and ethnic dishes raised the image of an everyday staple into a product for which some consumers are prepared to pay a premium. In the industrial sector, the use of rape seed oil as a substitute fuel for diesel engines substantially changed its perceived value; the ability of milk producers to deliver narrowly defined constituents for different processing uses substantially raised the value to the buyer.

CHANGING CUSTOMER PERCEPTIONS

Improvements in product attributes invariably require matching efforts to change customer perceptions of the new product offer. This requires promotional activity, which may be all that is required to exploit the benefits of an existing product that is basically sound: for instance, the 'naturalness' of wool and linen has been promoted in competition with synthetic fibres. The scope for this is widely underestimated, however, as Box 7.3 and Plate 15 suggest. In an alternative enterprise like accommodation or catering and leisure services, farm businesses also have a head-start in the ability to exploit the 'traditional' and 'nostalgic' image which some customer segments value.

PRODUCT STANDARDISATION

The most obvious area for marketing improvement in the agrifood sector lies in product standardisation, which combines a change of product with a change

> ## Staff of life
>
> Woeful. That is the word that adman Frank Bailey used to describe the part currently played by advertising and marketing in the promotion of bread or wheat-based products ...
> 'When the cheapest 100g of own-label bread is 3.4p against 15.9p for the cheapest 100g of cornflakes something's wrong. Surely the difference in cost of growing and processing the two staples has long been eroded. So it must be that people are prepared to pay this differential. But why? The answer is in the marketing which has installed brand values in cornflakes and supported the price increase ... This has not happened to bread, so the humble cornflake has been able to take off, leaving bread way behind.'
> In contrast to wheat lines, corn products have offered consumers added value and choice. Despite some product developments like the introduction of granary, ciabatta and focaccia, bread has largely remained, well, bread.
> 'This is not the case in Europe, and especially Germany, where, far from being in decline, bread is seen to offer variety, in the types and the different occasions you can use it.'
> So what could we do in the UK to boost bread's prospects? A simple variety message demonstrating that bread can be different could begin to awaken interest in an otherwise boring sector ... This message could be adjusted as necessary to target consumers more specifically ... the story behind this being that bread has all the natural ingredients you need for health. Sex and bread aren't out of the question either ... You see the couple having fun with ice-cream in the Haagen-Dazs adverts, and there's no reason why bread can't be made to be sexy as well, it just takes the will.
> 'Bread shouldn't be seen as just a commodity but as something everyone must have.'
>
> *The Grocer, 12 October 1996*

Box 7.3

of customer perception. Variability is a major problem of most agricultural output, which is not produced in controllable factory conditions. In some cases, particularly meat, what determines consistent quality and different quality attributes is not even understood. The result is that consumers buying food and processors buying industrial inputs cannot be confident that a repeat purchase will have exactly the same characteristics as the first.

The industrial user may sometimes be able to modify the manufacturing process to allow for minor quality differences provided the quality is known (in feed compounding, sausage or butter manufacture, for instance); sometimes different qualities can be used to make different consumer products. Where this is not possible, processors increasingly impose their own quality specifications which form the basis for acceptance or rejection. The same situation is familiar to potential suppliers of multiple retailers who, in the absence of other standards, have established their own inflexible standards of acceptance, which are then communicated at consumer level via their own brand assurance.

Clearly defined product standards of any kind have the additional advantage for both retailers and processors that they allow unseen purchase by specification or description. Purchase by description has been the norm in international commodities trading for many years, and it has made rapid advances in some agrifood markets following the introduction and use of computerised trading and electronic auctions (Chapter 9). Its advantage to the buyer is that it substantially reduces procurement costs by eliminating frequent visits to production units, marts, abattoirs, packhouse and processing plants, and the saving may be reflected in the producer price. The advantage to suppliers is that remote location is no longer a barrier to supplying the market. Provided the quality they can supply can be accurately described, remote producers can therefore compete on more equal terms with more optimally-located suppliers.

Quality standards

Compulsory minimum standards exist in a number of agricultural product areas, and products which fall below them cannot be sold or used for particular purposes. These represent the minimum core product which producers must supply if they are to find a market. Product standardisation has much more to offer, however, because it overcomes the inherent inconsistency of much farm output and permits product identification and assurance, together with different levels of product augmentation.

Some quality attributes vary in a continuous manner, and these can themselves be used as a basis for quality specifications: for example, the moisture content of grain, which allows purchase by content linked to a pricing schedule. In other cases a commodity is divided into bands, with the quality attributes of each band clearly defined and held constant over a long period. Examples of this are the EU horticultural and meat classification grades. (Strictly speaking, where the bands simply indicate *different* quality the process is called classification; where they signify that one is better than another, as measured by value, it is called grading. However, this distinction has broken down in many countries.)

Quality bands are normally defined by the quantity of the defining attributes present, together with a system of trade-offs. In some schemes a single attribute forms the basis of the assessment: for example, percentage of saleable meat for EU carcass classification (Figure 7.4), or weight in the case of eggs. Other schemes have a number of different parameters: for instance, size, ripeness and damage are components of the EU horticultural grading scheme.

The critical factor is that the attributes used must be measurable objec-

FAT CLASS
Increasing fatness →

CONFORMATION CLASS (Improving conformation)	2	3	4L	4H	5L
U+	74.4	73.3	72.0	70.6	68.4
-U	74.0	72.9	71.6	70.2	68.0
R	73.5	72.4	71.1	69.7	67.5
O+	72.8	71.7	70.4	69.0	66.8
-O	71.8	70.7	69.4	68.0	65.8

Figure 7.4 Saleable meat yields of EU beef grades

tively, with a consistent degree of accuracy. Ideally, they should also be attributes which are relevant to the consumer, though this is not always possible. For example, although the tenderness of meat is a key criterion for most consumers, it plays no part in meat classification, which is based on the yield of saleable meat. Similarly, when they have the choice, consumers freely select fruit which is riper than the highest EU grade. In both cases the explanation is that the system was devised for intermediary buyers, not consumers, and the quality attributes reflect their priorities (here, the return from a meat carcass, and the shelf life of fruit). They are *wholesale*, not retail standards, although they are sometimes used as such on retailers' shelves, with the result that consumer satisfaction is not best served by the classification scheme.

Meeting quality standards

Once quality standards have been established a farmer can adjust production to supply the needs of given market segments. If the standards are correct and reflect customer requirements accurately (intermediaries or consumers), a range of prices will emerge which reflects these needs. Buyers will sometimes publish these prices as an incentive for producers to aim for the higher level, and this strategy is very effective. In the 1980s, for example, the introduction by the UK milk marketing boards of price penalties and premia based on total bacterial count achieved a significant improvement in milk quality which persuasion had failed to achieve, and many producers maintained was impossible. Since then, milk production standards and supply specs have risen consistently higher, permitting very precise ingredients purchasing, and returning proportionately higher prices for a better-targeted product.

Success in the milk sector reflected the fact that the organisation setting the standard also decided the level of payment. Other schemes where this is not true (notably meat) have been less successful, which explains the proliferation of retailer and marketing group standards which impose their own quality control. Many of these specify husbandry practices and animal welfare (input) requirements as well as the quality characteristics of the product (for example, most multiple retailers' buying standards, milk buyers' supply contracts, and Soil Association/UKROFS standards for organic produce).

The saying that 'quality products sell themselves' is therefore increasingly true of agricultural output, because customers in a buyer's market, with a choice between different qualities of product at roughly similar prices, will always buy the better quality. This does not mean that a premium for quality necessarily exists – nor even, where a premium is paid, that it is sufficient to cover the extra production costs. It is also worth remembering that it is not always the highest possible quality that is required, but consistent quality. The important criterion is value for money, which is a combination of price and quality factors. This is true of both consumer and intermediary markets, where the perception and definition of value and quality depend on the use to which a product is put. For example, different qualities of apple are required for the fresh eating market and the juice processing market, and within each of these markets different buyers require different qualities and pay different prices.

NEW PRODUCT DEVELOPMENT

In most businesses, improving the quality of existing output to supply different market requirements will often improve profitability sufficiently to meet income targets and eliminate the need for more drastic action. However, a time invariably comes when it is necessary or appropriate to introduce a new product or enterprise because an existing product is reaching the end of its profitable life cycle, or additional revenue is needed. Changes to farming systems may also be required for biological or environmental reasons (disease or pest build-up, climatic change).

New product development (commonly abbreviated to NPD) is broadly the same whether the product/enterprise to be introduced is genuinely new (original) or simply new to the business. Most new products are in fact new only to a business, and not to the market. In agriculture, most introductions are transferred from other countries or regions where they have been fully developed and successfully grown (maize, oilseed rape, lupins), and all that is involved is a transfer of technology and known practices.

In the food sector many new products derive from other countries and cultures, and are simply manufactured from domestic ingredients (yogurt, fromage frais). Many others are 'me-too products': simple copies of other manufacturers' existing offers. New product development is not, in other words, dependent on an inventive mind. In the agrifood sector there has been relatively little genuine invention of new products, although this is changing as technology provides acceptable alternatives to traditional agricultural products. Mention was made earlier of the substitution of alternative protein sources for meat in ready-processed meals, which provides opportunities for crop growers at the possible expense of livestock producers (Box 4.5). However, innovation is most evident in the services added to existing food products, which generate most of the added value: for instance, new meat cuts which have added variety, convenience, and better value for money (Plate 16).

The sources of new product ideas are almost limitless, but may be:

- *knowledge-led*: ideas derived from new scientific principles or discoveries (for example, genetically-engineered foods); advances in industrial processes and technology, household technology (freeze drying, controlled atmosphere packing, microwave cooking)
- *demand-led*: changes in consumer habits, preferences, and lifestyles: new recreational expectations, attitudes to food, countryside and natural environment
- *environment-linked*: the need to cope with changes in the business environment – for instance, substitution of domestic for imported products, energy-reducing forms of production/processing; electoral pressures on politicians
- *supply-driven*: access to new or increased volumes of raw materials (new uses for surplus wool and arable crops, or a shortage of others).

In some companies a systematic trawl of the market for such opportunities is built into the management system; for the owner-manager, scanning the marketing environment and effective customer research serve the same purpose.

THE NEW PRODUCT DEVELOPMENT PROCESS

The new product development process is outlined in Figure 7.5. Many ideas will never get beyond the generation stage, but a good one can be considered at a rudimentary level before being subjected to more thorough investigation, which may involve experimental or trial production and extensive costing and market assessment. If the idea can be tested on a micro scale this is often the best – as well as the preferred – way of proceeding. For instance, a new meal ingredient could be grown on a trial plot and various cooking methods tried and preliminary consumer testing carried out; a prototype machine may be built and trialled. The development cost to this point will be fairly small (typically something less than 10 per cent of the final development cost), so relatively little will be lost if the idea fails. Costs subsequently mount steadily as the idea is turned into a marketable product, entailing all the supporting work of promotion and distribution to achieve acceptance.

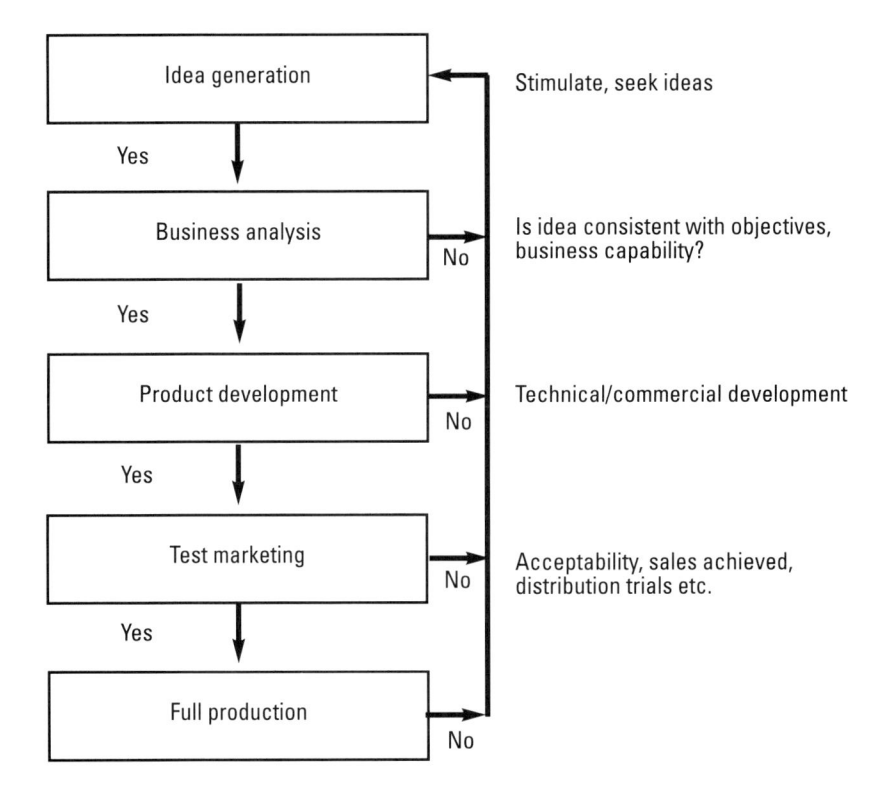

Figure 7.5 New product development process

The actual development stage is often lengthy and may be expensive if capital expenditure is required. Kitchen or workshop techniques must be converted into factory-style production, even for a relatively small-scale operation. This may entail modifications to the original product concept (for example, new ingredients better suited to the larger scale), and an entire product re-design may sometimes be required following product tests. Market investigation is simultaneously required to establish potential distribution channels, to identify appropriate and willing agents, to solve all the packaging and labelling requirements, and to determine pricing and promotion policies. Realistic cost and sales estimates must also be made to determine potential profitability.

By the end of this stage some 40 per cent of total development cost has generally been incurred, but little capital expenditure will normally have been committed. Proceeding to full production will increase costs rapidly and dramatically, so a test marketing at this point is usually desirable for a consumer product which is fairly inexpensive to supply and for the consumer to buy. For specialist products for which dedicated manufacturing capacity or facilities must be provided this is not possible. The only way to test the market for aerosol canned cream, for instance, was to produce the product first, which required substantial capital investment in manufacturing plant. A decision to proceed at the end of the development process may therefore have to be taken without market testing, on the basis of the appraisal and profitability projections. This is certainly true of most diversification enterprises which require substantial capital investment – which is why the systematic appraisal recommended in Chapter 3 is necessary.

Even where testing is possible it may be necessary to commit some capital expenditure for new plant or machinery, packaging materials, and some preliminary promotion, simply to produce test products. If at this stage the product's success is shown to be doubtful, it is possible in principle to decide not to proceed. In practice this option is receding as more sophisticated manufacturing techniques and integrated component/ingredient delivery systems are used, even at the level of the small-scale food processor. It also becomes less feasible as product life cycles shorten, since the need to capitalise immediately on a successful test marketing makes the lengthy process of establishing a production and distribution base, and building stocks of raw and marketable products, unacceptably risky.

For many products the latter stages of development and test marketing overlap, so a rapid launch is possible. At this commercialisation stage the highest levels of expenditure are incurred: up to 50 per cent of new product costs, most of it in capital expenditure and promotional spending. The greatest problem for small firms also arises at this stage, when access to retail stores and communication with target customers become necessary. Sometimes there may be no existing marketing channel for a new product. Farmers who have introduced some new raw materials have discovered that no marketing channels existed for the product, and a distribution system had to be created. In the UK, for instance, producer cooperatives had to be created to market rape seed and develop a wholesaling system for organic vegetables.

The key to success therefore lies in transforming a viable new product idea into a profitable business. Systematic analysis of opportunities as recommended in this book will be the necessary starting point, but the final decision will invariably rest on the subjective judgement of the entrepreneur.

CONCLUSION

The product decision involves the detailed design of a product with distinctive attributes which match the requirements of a well-defined market. In a strongly competitive marketplace of branded products, the aim is to establish a clear product identity which guarantees consistent quality and encourages repeat purchase, and allows promotion of the product's functional and psychological attributes. For raw materials, quality improvement coupled with labelling may be the best that can be achieved, possibly augmented by post-production services generally linked to quality assurance schemes.

CHAPTER 8

The price decision

The price obtained for a product has obvious relevance for enterprise viability, since price and numbers sold determine revenue received. The two factors are also interrelated because, other things being equal, the higher the price, the fewer products are sold. A business seeking to increase sales volume may therefore choose to reduce its price, and the only way a large supply is likely to be sold is to allow its price to fall. A business seeking to maximise profits rather than sales may choose to market a lower volume at a premium price.

Price also carries information about the quality of the product offered, so it must be consistent with the rest of the marketing mix: a high price is generally taken to indicate higher quality and added value, and a low price may raise doubts about product quality. Price judgements are also made in relation to benefits gained: *value for money*. The price customers will pay for a product is consequently a measure of the value they put on the benefits it endows, so the pricing decision involves judgements about customer psychology as well as production costs and competition.

The pricing decision is in other words complex, yet it is one over which a business has much less control than over product, promotion and distribution. This is true of all businesses, but even more true of farm businesses. The pricing function also varies significantly between raw product and consumer product markets. This chapter outlines the pricing function in competitive markets, and the opportunity for price risk management which is available to producers of some agricultural commodities via the futures markets. These are a normal feature of agricultural marketing in many countries, and are likely to become so everywhere as trade liberalisation progresses.

Essentially, two issues are involved:

- *price level*: how is or can the general price level for a product be determined?
- *variation around the price level*, reflecting quality/quantity purchased, time/place of sale, use values, distribution costs (for example, added delivery costs from/to remote regions).

In some circumstances a business may succeed in being a price *maker*, deciding the price it wishes to charge for a product and actually realising this price. This is usually possible only in the short run; in the long term competition will determine the price achieved. A business will thus normally be a price *taker*, obtaining the price set by the market. Once a customer is in a retail store, for instance, the price requested must be paid in order to acquire a given product: the retailer is the price maker and the customer is the price taker. The situation is nevertheless short-lived, for the customer may take his business next time to another store with lower prices, and if many customers do this the high-price store will have to lower its prices to stay in business.

In the real world many more factors than this are involved, but the example illustrates the basic principle of price theory: that the price of *identical* (undifferentiated) products will normally fall to that accepted by the lowest-price seller. This assumes three conditions: that products are perceived as identical by the customer; that customers have sufficient information about products and prices elsewhere, and that they understand and are able to allow for different transaction costs in different localities and at different times.

In practice competition rarely produces immediate price falls to the lower level, and prices for identical products may not equalise because the market fails to achieve any or all of these conditions. Most importantly, customers may not perceive products as identical, and price differences arise as the result of different customer evaluations of product, promotion and distribution. Attempts to raise prices are thus more likely to succeed if they involve the manipulation of one or more of these variables to change the customer's value perception and the price he is prepared to pay.

This is equally true of raw material and end-use markets, since processor prices for raw materials reflect end-use value. Many raw materials have multiple uses, and the different prices they achieve determine product allocation, with the highest value use commanding supplies and the lowest value uses possibly failing to obtain supplies. These *derived prices* are what farmers receive for commodities, and they are rarely known until after the production cycle is completed. They may therefore vary considerably in relation to changes in supply and demand which it is often impossible to forecast, or only when it is too late to change the underlying situation.

It is the interactions of price and revenue, revenue and income, and price and supply which have caused governments to intervene in agricultural markets (and in consumer food prices in most developing countries). In some countries efforts are also made to achieve lower food prices through the subsidisation of farm inputs, with the result that agricultural prices are overlaid with a political significance not usually associated with other commodities.

PRICING OBJECTIVES

The overall pricing objective is to ensure that sufficient sales are made to allow the business to achieve its business objectives: pricing objectives therefore depend on business objectives. In the long run prices are determined by the market, but in the short term a *price-making* business may be able to set prices to satisfy its objectives, and even price *takers* will know the prices they need to satisfy their objectives. Four broad pricing objectives are distinguished: business survival, sales maximisation, current profit maximisation and product quality leadership.

BUSINESS SURVIVAL

Over the long run business survival depends on revenue covering costs (including a level of profit sufficient to reward capital providers), and revenue should ideally exceed cost sufficiently to provide a surplus for reinvestment. In the short run, most businesses experience periods in which reduced demand or overcapacity in the industry may cause revenue to fall and prevent total recovery of costs. For businesses subject to seasonal supply or demand this is almost inevitable, and highly likely for businesses exposed to the fluctuations of the international trade cycle.

Provided these periods of deficit trading are short, and revenue gained exceeds the variable costs associated with production, production may continue since any surplus will make some contribution to the fixed costs (which would be totally lost if production ceased). A business may in other words survive a period of unprofitable trading provided it succeeds in the long run in covering its fixed costs, but the ability to do this may depend on past performance and the attitude of capital providers. It is only the insecure business in which deficit trading will inevitably lead to bankruptcy, and in this respect the family business is often better placed to survive than a shareholder company dependent on market assessments.

The security of the business also dictates the pricing strategy to be adopted in a deficit trading situation: whether to reduce prices in the hope of generating more sales, or keep prices at current levels and accept lower sales volumes, or to cut costs by short-time working, lower staff levels, increased efficiency.

Any one or a combination of these strategies may be followed by a secure business, but not one whose capital providers are uncertain about its future. This leads to the most commonly stated long-run pricing objective of businesses: *to achieve a target rate of return on capital invested.* This is commonly set at between 8 and 20 per cent after tax, which should meet a reasonable income target and provide a return to capital providers sufficient to prevent them selling shares or foreclosing loans.

SALES MAXIMISATION

Sales maximisation (or at least the objective of obtaining a high market share for a product) is a common strategy, supported by strongly competitive pricing. Large volumes permit lower production costs by achieving economies of scale; they also improve the efficiency of other marketing costs (principally promotion) since the costs are spread over a much larger volume.

As a long-run strategy, sales maximisation leading to a dominant market share may allow a business to raise prices and gain extra profits in the absence of little or no competition. High profits will invariably attract competition, however, and some customers may be alienated if they feel exploited. Large companies or business sectors will find that governments intervene under anti-trust and competition law to prevent them from achieving this.

CURRENT PROFIT MAXIMISATION

Economic theory states that profit is maximised at the point where the cost of the last increment of production (marginal cost) equals revenue derived from the sale of this last increment (marginal revenue). In a competitive market this marginal revenue is the price. For the producer of a non-differentiated commodity sold in a highly competitive market prices do tend to approximate marginal cost, and the producer has little control over the price received. Profit maximisation must therefore be sought via cost reduction rather than a price change. A multi-product business with differentiated products in a less competitive market may have some control over its own prices, which may allow an increase in profit level to be achieved. The manager considering this will need to take account of profit per unit *and* overall profit level (a function of numbers sold), and in some situations low unit profit is acceptable provided this encourages large sales volumes.

Sales maximisation and profit maximisation objectives may cause conflicts within the same business. A business with employed managers may find sales maximisation pursued at the expense of profits if management remuneration is based on turnover or throughput, not profitability. In the owner-managed business it may seem logical to maximise profits since profit = income for the owner. In practice many businesses need to achieve a satisfactory compromise between the two objectives, a common solution being to pursue sales maximisation subject to maintaining/achieving a satisfactory (stated) profit level.

PRODUCT QUALITY LEADERSHIP

Since many consumers associate higher prices with higher quality, a business may deliberately choose to charge a high price to establish a quality image. Quality may not come cheap, however, so the costs of production, inspection and quality control must not increase disproportionately to the additional price that may be charged. In many cases the price that may be charged is nevertheless significantly higher than can be justified by increased costs, and it is imposed simply to reinforce the consumer's perception of a product's exclusiveness (fine wines, limited production of a special/rare crop variety, up-market restaurants, designer items).

Innovative products enjoy the same special character which may command a higher price, even though they are not necessarily of higher intrinsic quality. In this case the novelty *is* the special quality. Sometimes the premium is justifiable in terms of the R&D costs involved in new product development and marketing, but it is the customer's perception, not production costs, which command a premium. The same may be true of niche products made by small specialist producers (for example, farmhouse dairy products, the special character of farmhouse accommodation). The price charged should therefore reflect this status rather than the cost-plus formula adopted by many innovative entrepreneurs.

PRICE LEVEL

The price level of a product will be the result of a number of interacting factors shown in Figure 8.1. Their relative importance will vary from product to product and over time. For raw materials like feedwheat, prices are likely to be the result of forces of supply and demand on a given market on a given day. For an innovative new product a business is likely to charge as much as it can in the short run, knowing this premium will be eroded as the offer is matched by competition (consequently, the longer this can be delayed by patents, copyrights, and licences, the better). For many products in the mature stage of the life cycle, the price is likely to be based on the cost of production plus a rate of return sufficient to keep the firm in business and a margin sufficient for reinvestment in plant, machinery and new products.

Four main approaches to price level determination are distinguishable: competition-oriented, cost-oriented, profit-oriented and demand-oriented.

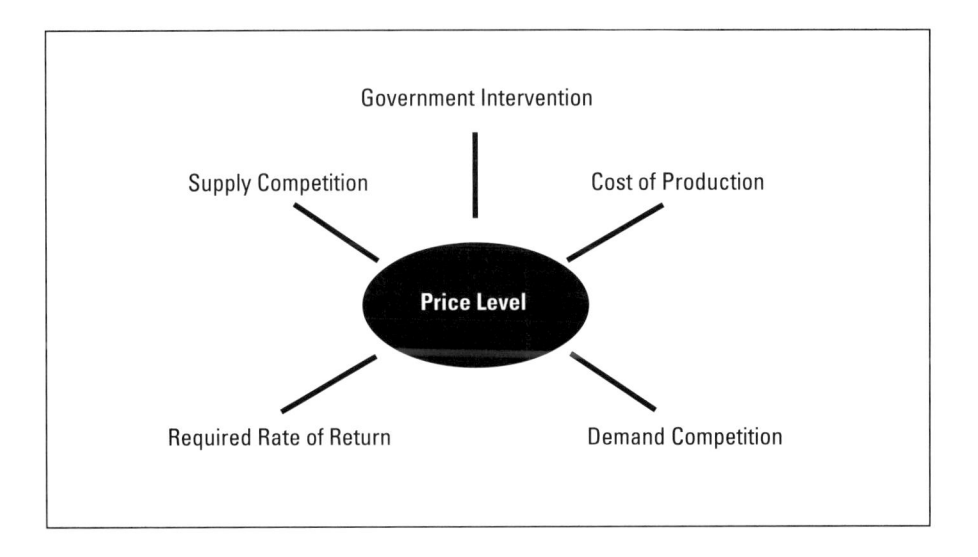

Figure 8.1 Factors affecting price determination

COMPETITION-ORIENTED PRICING

In competitive markets, the effect of price changes on future production and of production changes on prices can have very damaging effects, depending on the reactions to a current situation by a large number of independent, unorganised entrepreneurs. A basic understanding of the pricing mechanism in competitive markets is therefore essential.

The price received by most primary producers is the *equilibrium price*, determined by supply and demand in the marketplace. In Figure 8.2, price P = the supply and demand for the commodity in terms of quantity Q. At a

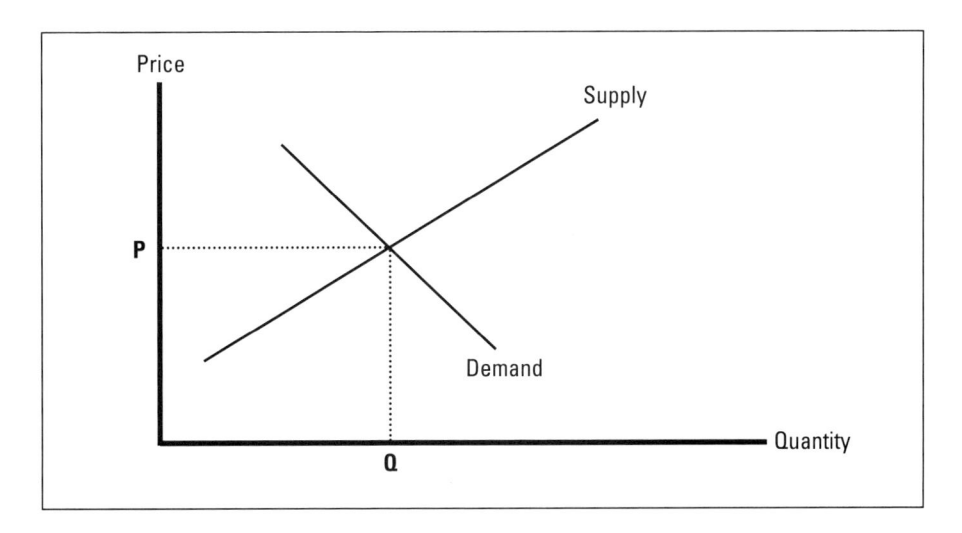

Figure 8.2 Equilibrium price determined by supply and demand

higher price supply would increase but demand would fall; at a lower price demand would increase but supply would fall.

This is true of the market as a whole, and its implication is that producers can sell as much as they like at the equilibrium (market-clearing) price, but at a higher price they will sell nothing. (In technical terms, the producer faces a perfectly elastic (horizontal) demand curve.) This is termed a perfectly competitive market, which satisfies the following criteria: undifferentiated products, perfect customer information, multiple suppliers powerless to affect the price, free market entry and exit of suppliers.

The agricultural industry and its raw material markets exhibit these characteristics, although at any one time and place a perfect market is unlikely to exist because the maximum level of supply is limited by the impossibility of altering production and/or the difficulty of bringing product immediately onto the market. Price will therefore be determined by the demand for that fixed level of supply: if demand is high, the price will rise, and if it is low, vice versa.

The effect of a change in price on demand or on supply (elasticity) may be calculated by the slope of the demand and supply curve, defined as:

$$\text{price elasticity of demand} = \frac{\text{percentage change in quantity demanded}}{\text{percentage change in price}}$$

$$\text{price elasticity of supply} = \frac{\text{percentage change in quantity supplied}}{\text{percentage change in price}}$$

Elasticity figures are published for many commodities which show that for necessities and products with few substitutes demand is inelastic for reasonably small price changes. Where substitutes are plentiful and substitution is easy, demand is elastic. Product differentiation (for instance, branding) reduces elasticity because it tends to reduce consumer willingness to substitute one product for another; this allows prices to rise, with consequent benefits to revenue. The demand elasticity for food is very low in all developed countries, for the obvious reason that the population is reasonably well fed; consumers would therefore not purchase substantially more however low the price of food falls. By contrast, the effect of price falls on holidays and recreation is significant: demand is said to be *elastic* whereas for food it is *inelastic*.

Using published elasticity figures, the effect of changes in supply both on price and on total revenue derived from the sale can be calculated. The question is often asked, for example, is it better to sell a few at a high price or a lot at a lower price? If 'better' means higher revenue, the question can only be answered if the elasticity of demand for the product is known. In Figure 8.3, for example:

ELASTIC			**INELASTIC**		
Price	Sales	Revenue	Price	Sales	Revenue
2.00	2,850	5,700	1.00	4,770	4,770
3.00	930	2,790	1.50	3,810	5,715
Price increase → revenue fall			Price increase → revenue increase		

THE PRICE DECISION

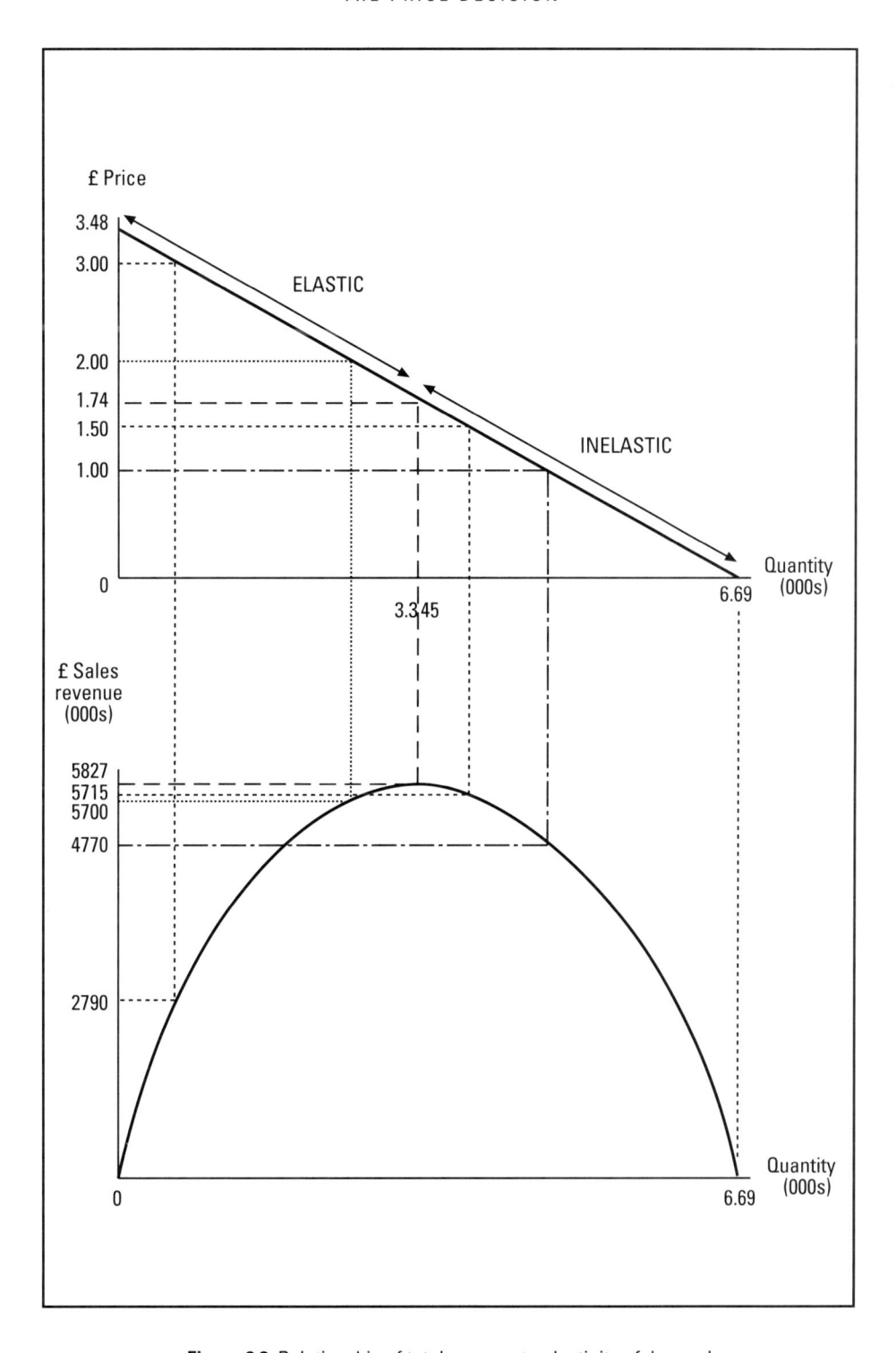

Figure 8.3 Relationship of total revenue to elasticity of demand

121

Consequently, a supplier who can achieve inelastic demand for a product may raise prices and increase revenue. Farmers will be familiar with this concept since many support systems (quotas, for example) restrict the quantity marketed in order to raise prices, thereby increasing total revenue.

Any form of disequilibrium in the market will prompt a reaction both by buyers and sellers. A shortage of supply may encourage producers to expand production or search out additional supplies to trade on the market. Customers will respond to shortages and high prices by seeking supplies in other markets or substituting other products. In the long run supply and demand competition will establish the price – where 'long-run' is defined as the time taken for supply and demand conditions to change. How long this takes will depend on the complexity of the market and the need for research development, investment and implementation by market participants. Chronic market disequilibrium clearly requires a more vigorous and sustained response than a short-term interruption of supplies or downturn in demand.

The problem for the manager is to forecast future prices on the basis of today's. This means accurately recognising and assessing the cause of any unusual price existing in the current market. If it has a physical cause (for example, drought-related crop shortage) it may not recur and normal production may reasonably proceed. If abnormally high prices are caused by some underlying change in market conditions which is not identified and production proceeds at normal levels, lower prices may result if the assumption was wrong.

The greater the number of producers making uncoordinated decisions, the higher the likelihood of 'booms and busts' resulting from these cycles. The cycles are characteristic of agricultural commodity markets, especially enterprises with relatively short production cycles (chickens, eggs, pigs). The cyclic effect is not related to the products themselves, however: it reflects the fact that supply is determined by large numbers of relatively small, uncoordinated suppliers, which prevents production and marketing planning on a rational basis.

The conclusion drawn by some producers is that trading on such markets should be avoided. The alternative is the introduction of systems which allow some planning for the uncertainty. In agriculture this has been achieved by government intervention, but the risk may also be reduced by cooperative marketing, through producer-buyer alliances, and by private futures trading. However, none of these will over-ride the fact that the price level will eventually be set by the interaction of supply and demand in the market.

COST-BASED PRICING

All businesses seek to relate prices received to cost incurred, but sellers of undifferentiated products are relatively powerless to influence the price they receive. Producers of other products may in the short run be able to charge a price based on costs, but as explained earlier, lack of demand or competitive pressure may force them to reduce prices below their own cost-based price level. The fortunate producer may be able to raise prices above that strictly necessary to recover costs.

Cost-based pricing depends on an accurate measure of costs, which any business might be expected to be able to determine fairly readily. In practice this is rarely true. A business producing a single product sold in a single size by a single method could charge all the business costs to that product, but this is so rare as to be unrealistic. Normally business costs have to be allocated to different activities associated with different products, which causes difficult pricing decisions. Four types of cost have to be allocated to the production of an individual product.

- *Variable costs* vary directly with the amount produced (raw materials, packaging, piecework remunerated labour, etc.) and are easy to identify and allocate.

- *Fixed production costs* do not vary directly with output, and include costs associated with having and running plant and machinery. These can be allocated fairly accurately to an individual product, but the cost per unit will fall as production expands.

- *Fixed cost overheads* are associated with the existence of the business itself, and include R&D, public relations, premises, owner/director salaries, etc. Since most attempts to find a proper allocation procedure are unsuccessful, arbitrary allocations are generally made.

- *Profit* is a cost, since it is necessary to remunerate risk capital and entrepreneurship; it must therefore be included in any cost-based pricing formula. Unlike some other costs (including wages) profit does not have to be paid out on an ongoing basis, but it is usually paid periodically, and in difficult times may not be paid at all. In this sense it is a discretionary cost.

Overheads and profit are extremely difficult to allocate to individual products. Various methods based on proportions of total turnover are employed, but none are completely satisfactory and certainly do not provide a basis for pricing. Given nothing better, allocation by proportion is the only solution, and once the allocation decision is made the price necessary to cover all the costs may be calculated.

The cost-plus method has the advantage that customers understand it and generally see it as fair. Its major weakness is that it disregards the level of demand, and may lead to a price at which demand is zero or at least very low: production would therefore result in unrecoverable losses. Conversely, if customers would pay more than is being charged, demand will exceed supply, and a revenue opportunity would be missed. Supplying this demand may entail increased investment in additional capacity, but this could be funded by the increased profits from higher prices and volume. The alternative is to 'ration' the limited supplies, but this could encourage trading in the product by third parties who buy and sell on at a higher price. If a business sells a successful product too cheaply, it also encourages other suppliers to enter the market with a competitor product at a higher price.

TARGET-LEVEL PROFIT PRICING

Target-level profit pricing (the commonest method) is a more sophisticated variant of cost-plus pricing, which uses the economic concept of elasticity of demand to calculate changes in demand resulting from price changes. This 'break-even' method is illustrated in Figure 8.4. The fixed cost

(£6,000) and variable costs (£5/unit) are depicted in the normal way so that their vertical summation creates the total cost line. The total revenue line is the product of price and quantity (TR=PxQ). The slope of the line is equivalent to the price. In Figure 8.4, if the price is £15 per unit to cover costs, sales of 600 units are necessary to break even (cover total cost without profit: TR=£15 x 600 = £9,000). If a profit of £2,000 is required with a price of £15, the sales figure would need to be 800.

Total revenue = £15 x 800   £12,000
Total cost = £6,000 + (800 x £5)   £10,000
Profit   £2,000

If it were desirable to sell less (or production constraints made this necessary), the price would need to be increased. Similarly, if higher profits were required at the same level of output, a higher price would be required, which would be represented by another line on the diagram with a steeper slope. A higher price will result in fewer sales, but the magnitude of the fall will be determined by the elasticity of demand. In the example above, if demand is known to be inelastic, the price could be raised to (say) £20 per unit, sales would fall from 800 to 600, but revenue would remain the same at £12,000, and profits would rise to £3,000.

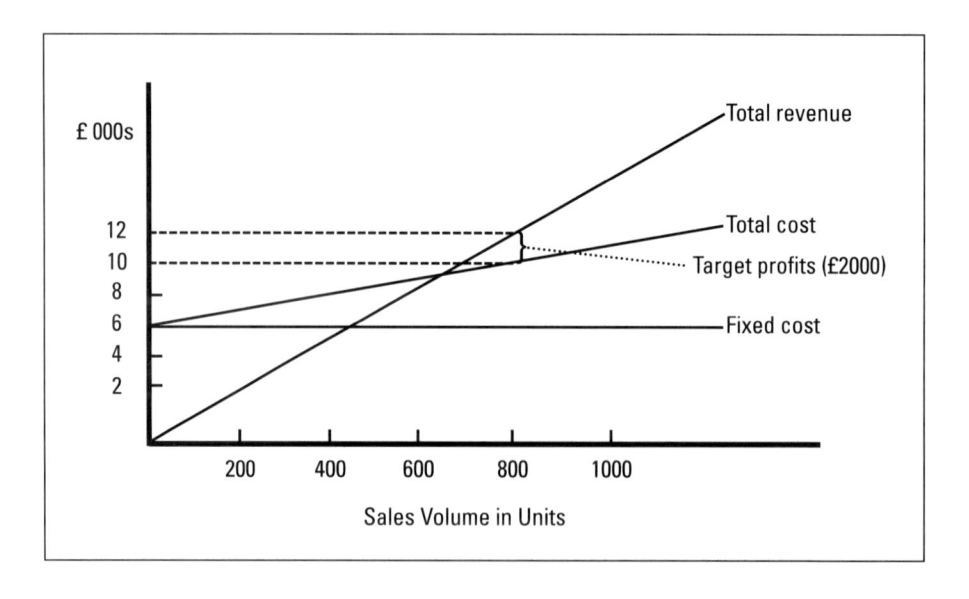

Figure 8.4 Break-even chart for determining target price

DEMAND (CUSTOMER)-ORIENTED PRICING

Break-even pricing acknowledges that demand must have a bearing on the price asked, but it does not allow for customer perceptions reflected in the adage 'charge what the market will bear' (if demand is strong charge a higher price than when it is slack). Customer perceptions and reactions are reflected in two types of pricing method:

- *price discrimination* reflects customer attributes
- *perceived value pricing* reflects customer perceptions of product attributes.

Price discrimination

Price discrimination charges different prices to different customers for the same product. The obvious example is the differential pricing of raw materials which reflects end-use values, but the principle is valid in most business situations.

Different prices may reflect cost differences associated with production or marketing (for instance, quantity discounts, reflecting reduced labour and packaging costs). The objective of this may be to:

- even out demand peaks and troughs: for example, senior citizen and student discounts on slack trading days; bargain break weekends and seasonal discounts in the holiday/leisure sectors to even out and improve year-round throughput
- generate throughput by special incentives to some customer groups: family discounts for leisure facilities; 'singles' or 'standby' prices for accommodation.

The price differentials may not necessarily be evident in different marked or published prices; indeed, this may be unwise because it may annoy customers unable to benefit, and alternative methods will achieve the same objective: for example, a discount card; a loyalty bonus payable in kind (free tea in the farm shop); money-off another product.

Perceived value pricing

Since customers purchase psychological benefits as well as intrinsic product attributes, they may pay a higher price for a product which maximises their perceived value. The perception may be based on a tangible product improvement (taste and degree of ripeness of fruit; mature cheese; 'crispier' cornflakes) or solely on the customer's perception. The latter is often associated with a brand name (Kellogg's cornflakes), and the higher price of the branded product is part of the benefit it confers (if it is more expensive, it must be better, or it shows off purchasing power). Conversely, if a product is cheaper some customers will assume it is inferior. Good customer research will allow judgements to be made about such attitudes to price and product requirements, and competitor analysis will suggest the need and scope for price competitiveness.

VARIATION AROUND THE PRICE LEVEL

Once the price level for a product has been determined, the exact price to be charged under different circumstances must also be determined: at different times and places, and for different quantities, product forms and sizes within a range. Experience and analysis will be helpful in this, but an element of luck will always be involved which no formulae can supply. A decision once made is not cast in stone, moreover: price modulation ('shading') is usually possible, and often features prominently in the marketing strategy.

NEW PRODUCT PRICING

The pricing decision may be more difficult where a new product is introduced, because all the components arise simultaneously, and the producer has insufficient experience and detailed knowledge of consumer reaction both to the product and its price. The difficulty can be exaggerated, however, since it is exceedingly rare for a product to be introduced for which there is no price 'comparator' – albeit an imperfect one, and for most products price comparisons are relatively straightforward.

Two basic new product pricing strategies exist which are appropriate to certain business objectives and market situations: *price skimming* and *penetration pricing*.

Price skimming

Price skimming exploits market segmentation, and is most appropriate to markets and for products where innovation and novelty are highly valued: for example, luxury or state-of-the-art products, new restaurants, health farms. The method is implied by the analogy with skimming cream off milk – the richest fraction. The small sector of the market which is willing to pay a higher price is targeted first, and only when it has been satisfied is the price reduced to appeal to the next segment, and so on down the scale until the product price is low enough for everyone to buy it. In the meantime, to offset the price fall, new products are introduced to appeal to the top market segment.

Price skimming allows a business to achieve the maximum profit from each market segment. It also provides a basis for the allocation of new products in relatively short supply; for planned marketing of wider markets as production expands, and for economies of scale which permit lower prices to be charged as a result of lower costs. The time period for each price level will vary considerably between product groups. For some food products the duration of the 'cream' market may be very short indeed, particularly where me-too copies are easily introduced. For some larger purchases like farmhouse accommodation there may also be a relatively short period, possibly only one season, in which a new offer can cream the market.

Penetration pricing

Many food market sectors are very congested (dairy products, biscuits, convenience meals), and the products are relatively inexpensive and easily copied. In these circumstances entry at a sufficient volume is difficult, but easier if lower prices are charged. This *penetration pricing* is the traditional method for fast-moving consumer goods, including foods. However, where a market is very competitive, prevailing prices may be returning only a modest profit to existing competitor products. If the newcomer needs to offer a price advantage to gain entry, zero profits or actual losses will therefore be incurred. This is unsustainable in the long run, and the business will have to anticipate and plan for price increases, which can arouse consumer resistance and complaints of exploitation which result in lost sales.

To avoid this, many businesses announce introductory price offers, the understanding being that a discount is offered for trial, but the long-run price will be higher. If sales are sluggish the introductory period may be extended, sometimes for a very long period; if take-off is rapid the discount period may be curtailed before the planned (unannounced) termination date. Some customers will still be lost following the introductory period, so this must be allowed for in the planning. (Food consumers are particularly wise to penetrative pricing, and switch brands regularly to exploit the practice.)

PRODUCT RANGE PRICES

Many products are part of a range (of qualities, sizes, varieties, models, etc.), and a range of prices has to be arrived at which reflects these differences. The basic principles of range pricing are clear and widely understood by buyers and sellers: the larger the pack size, the lower the unit price; the more basic the product, the lower the price; the higher the added value, or the newer/more exotic the variety, the higher the price. The details are much less clear, but provided these broad principles are obeyed, the actual product price will depend on competitor products in the same band of the range.

Problems may arise if pack sizes do not allow direct comparison of value for money, but in most developed countries *unit pricing* is a legal obligation. Quality remains problematical (Chapter 7), but claimed quality advantages must normally be seen to be real to achieve a price premium. Varietal prices rest on more subjective consumer evaluation. For example, some consumers will pay more for brown or white eggs, or an old fruit or vegetable variety, but why they do so (status or perceived better intrinsic value?), and at what level price resistance will be encountered, is a question for customer research.

COMPETITIVE PRICING

Price changes can occur at any time in a product's life cycle, and are actively used as part of the marketing mix in order:

- to generate sales
- to increase the competitive pressure on other products and suppliers
- as the basis for some marketing activity – for instance, promotion to raise product awareness, modify product image, etc.

In some markets where competition is very strong, price competitiveness may be a necessary condition to achieve sales. This is typical of geographical and commodity markets, where many customers buy a product irrespective of brand (petrol, milk, bread, flour, etc.). Continuous price shading therefore occurs as each business seeks to offer the lowest prices, and this may lead to a price war, with prices spiralling out of control. More commonly it leads to a multitude of minor changes which need careful management in the light of their financial implications and promotional activity, in which price shading is a major support.

Periodic sales

The most visible price reduction method is the periodic disposal sale during which prices are reduced on some or all products to reduce surplus or obsolete stocks. Many seasonal enterprises routinely employ such sales to dispose of unsold stock which cannot be carried to the next season; year-round enterprises need to make room for new stock. In times of recession there may be a continuous 'sale', and the sale price is effectively the normal price. For some businesses this is the normal way of doing business, whether or not a 'sale' is advertised. However, there are legal restraints on what may or may not be claimed about the level of price discounts relative to competitor prices and to previous prices charged by the same supplier.

Special prices

A normal feature of everyday trading is the 'special price' to encourage purchase. In a multi-product enterprise special prices may be used as a matter of course for different products within the portfolio, to raise or sustain the profile of the business and win customers away from other suppliers. The price reduction is thus the message for the promotion mix (Chapter 11), and may take many forms. Some are simple price discounts, possibly associated with a seasonal or other surplus; others are designed to stimulate additional sales: money-off next purchase; associated product offer (half-price cream with full-price strawberries).

These price promotions are very common, and are generally considered to have short-term benefits both in encouraging trial purchase and extending the period before customers need to re-purchase, thereby denying sales to competitors. In the retail store they are a normal part of product display and salesmanship, designed to attract attention to neglected items and shift slow-moving products.

Loss leaders

The 'loss leader' offers a highly publicised or essential product at a very low price in order to attract customers into a situation where they will buy other products at full (possibly inflated) prices. This is a normal part of competitive strategy in the retail food sector, where the loss leader is typically one of a small range of staple products which most consumers have to buy: milk, bread, some canned products such as baked beans. The price is generally very low – invariably below cost price (hence 'loss leader'), and the practice is one from which small businesses suffer, particularly specialist suppliers like bakers, fresh produce farm shops, whose loss of trade to a loss-leading supermarket can force closure. It is a practice which small businesses may also use, however, to attract customers to a relatively unfavoured location or in competition with other providers: for example, refreshments are commonly offered as a loss leader by leisure activity providers. The strategy nevertheless depends on actually realising additional sales from other products.

The weakness with all these price-led offers is that the shrewd customer shops around for the best buys. Many will go to several places to benefit from price offers without making other purchases, or take advantage of a special offer without making any additional purchases (a common situation where subsidised admission to leisure facilities is allowed as an incentive to further spending which is not realised). This may be acceptable in some circumstances, particularly to attract customers to a new offer. As a long-run strategy it may lead to volume at the expense of profit, and establish a low-price image which is difficult to modify.

PRICE VARIATION

Price variations reflecting quantity, place, and time are a normal part of trading. The first two can be relatively easily managed; the last is a particular problem for businesses which experience seasonal fluctuations in supply and demand.

Quantity-related prices

Bulk buying discounts (unit prices which vary according to quantity purchased) are part of normal terms of sale at all levels of business (low tariffs for volume users; 'three for the price of two', or 'an extra 10 per cent in the bottle'). These reflect the reduced cost of administration, order processing, lower packaging and delivery charges etc., but the discounts may be larger than accounted for by costs alone, simply to encourage larger orders. They may also expand consumption because they reduce the number of 'stock-outs', when a buyer allows the product to go out of stock.

'Over-riders' are extra quantity discounts demanded by larger customers simply for doing business with a firm and providing an outlet. A Monopolies & Mergers Commission investigation in the UK in the 1980s concluded these were not justifiable on grounds of reduced cost to the supplier, but represented the simple application of buyer power. They nevertheless continue to be demanded, and suppliers trading with a large buyer (processor or retailer) may find they have to give them. In some countries over-riders have to be given in the form of advance cash payments to the buyer, rather than resulting in reduced sales revenue.

Spatial Pricing

A seller will determine the price he wishes to charge at his production site, to which a margin must be added to allow for delivery and other distribution-related costs. Many industrial products and raw materials are priced ex-works, and if delivery is required it is charged for separately. In this situation it may be preferable to include the delivery cost and quote a delivered price, which may not actually vary much from region to region. Since cost does vary with distance, however, cost-pooling must be introduced, with shorter distances subsidising more remote delivery points. In practice, the decision will usually be made by the customer.

Seasonal/temporal pricing

Prices in competitive markets vary over time as a result of changes in supply and demand. In markets where firms are able to set their own prices, it is common to reduce prices when demand is slack in order to reduce stocks, rather than incur storage costs resulting from inability to reduce production. Where this is not possible products are stored and an attempt is made to recover storage cost via a price increase. Both policies are used by producers of consumer products who are price makers.

In commodity markets producers are essentially price takers, and their price is subject to large fluctuations in supply (and to a lesser extent also in demand) which result from predictable seasonal climatic factors and unpredictable natural events (drought, disease) and world trade conditions. For example, Plate 17a shows normal seasonal price variation over twelve months, including the predictable price fall when the new crop becomes available, and a dramatic price level change between the two years reflecting drought-related supply changes.

Seasonal price changes over the year are well-known to farmers, and the only decision required of them is how to respond. For example, most raw milk buyers publish a seasonal scale of payments up to a year in advance, and the scale is modified to encourage a production shift from a low-price (high supply) to a high-price (low supply) period. For storable commodities like grains, buyers offer higher prices in some months to encourage the storage and progressive release of supplies onto the market. For commodities sold at auction a seasonal price pattern has emerged over the years which achieves the same effect. UK produced lambs, for instance, command higher prices in spring-to-early summer when supplies are low, and low prices in the autumn when supplies have historically been high (Plate 17b).

Faced with such fairly predictable price changes the producer has only two basic marketing options:

- the production pattern may be shifted to benefit from higher prices at particular periods (for example, the calving pattern of part of a dairy herd may be changed to fill the July-August trough and reduce output at the May peak; out-of-season lambing is possible, though many producers have found that in this case the extra costs incurred are not recovered through increased prices, so the production system is not common)

- where production cannot be changed a product may be stored, providing the increase in price more than compensates for the costs of storage.

In both situations there is an expectation, based on past trading, that prices will be higher at the end of the storage/production period, and support arrangements often ensure that this is so. The intervention prices set by the EU, for instance, are seasonally adjusted to cover the costs of basic storage. Such seasonal schemes are still not immune to instability caused by external causes, however; the producer therefore remains vulnerable, and this vulnerability increases where fluctuations have no underlying seasonal pattern that is predictable. In both cases, futures markets offer a means of stabilising revenue and reducing the risk.

FUTURES MARKETS

In many countries, commodity-linked futures trading markets provide a way of insulating business revenue from price instability in the commodities traded (for instance, wheat, barley, potatoes, frozen orange concentrate). Whereas institutional (government) price stabilisation reduces price fluctuation by removing the causes and hence the effects of price changes, futures markets do not affect the prices realised on the market: they allow producers to protect themselves (*hedge*) against the effects of these price changes.

Most futures markets were founded to provide risk cover against price fluctuations during protracted storage and long-distance transportation. In the days before air freight, decisions to ship might be made on the basis of one price, but by the time the shipment reached the market the price could be significantly lower, resulting in a loss of revenue. Similarly, when a commodity was put into store in anticipation of a price rise which did not materialise, a loss would be incurred. To encourage long-distance shipping and storage a means of guarding against loss was required. Hedging against possible loss was therefore organised on a regulated futures market, which came into existence alongside but independently of long-established commodity markets where physical commodities are bought and sold. Futures markets are thus not a new phenomenon: they are almost as old as commodity marketing itself.

HOW FUTURES MARKETS WORK

On the commodity markets trading may be for immediate delivery or a forward sale for some future date, and physical trading takes place: lots of commodity physically change hands, and different specified characteristics are agreed between buyer and seller (quantity, quality, location and time of delivery etc.). Trading is between individuals who own the physical commodity and others who need it, and possession (title) is exchanged.

In the futures market trading is not in commodities, but in contracts: agreements to buy or sell a standard quantity and quality of a given commodity at a specified date in the future, at a price established when the contract is entered into. Physical exchange is neither intended nor expected. Physical exchange is technically possible (a fact of major theoretical importance) but it rarely occurs, and for many commodities it is not even permitted.

In futures contracts (unlike physical contracts) all the key variables – quantity, quality, delivery date – are determined by pre-established rules for each commodity contract and are non-negotiable. They are standardised contracts, and price is the only detail left for buyer and seller to determine, via their respective brokers. Futures trading is possible only within an organised, regulated and centralised marketplace, which ensures that prices are determined by visible, competitive bids, either between individuals physically present in the market who bid by open outcry, or electronically via computer screens in the marketplace and the offices of brokers. All trading must be undertaken by a broker acting on instructions from a principal (buyer or seller). For this service and the use of the clearing procedure the broker charges a commission negotiated between the two parties, which will partly depend on the scale of business undertaken.

The theory of futures markets trading is basically very simple. On any one day there are two prices for a commodity: the spot price (= today's price for physical delivery) and a futures price which is today's price for the commodity to be delivered at a later, specified, date. In principle the difference between the two is the cost that would be incurred for storing the commodity from today until the specified date (the *carrying charge* or *basis*) (Figure 8.5). As the delivery date approaches the carrying charge diminishes, and on the delivery date (date of contract maturity) it will be zero, and the two prices will be identical for commodity of similar specification.

The relationship between the two prices is ensured by the opportunity for traders to make a risk-free profit by buying in one market and selling in the other: this is known as *arbitrage*. For example, if one month before maturity the difference between the spot and the futures prices for wheat was £10/tonne (spot price £115 and futures price £125) and the storage cost was only £5, it would pay a trader to buy physical grain and put it into store. At the same time he would sell the grain on the futures market for £125, and on the date of maturity the grain would be sold or delivered against the contract, at a profit of £5/tonne.

Given freely available information about prices, it will be clear to many traders that a guaranteed profit can be made, and they will all buy physical

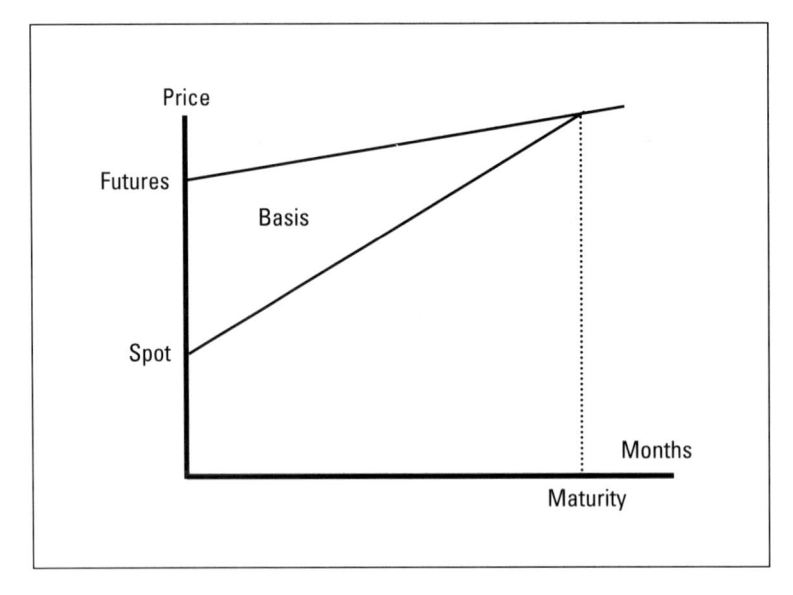

Figure 8.5 The relationship of spot to futures prices

grain on the spot market and sell futures contracts. Grain prices will be forced up and futures prices will fall, until the opportunity for profit is excluded, at the point when the difference between the two prices is simply the storage charge from the date of purchase to date of maturity.

This process ensures that major differences between the basis and the real cost of storage are quickly corrected. More important is the opportunity for arbitrage which ties the two prices together, with the result that a change in one provokes a change in the other. The market thus reflects the true supply and demand situation. If supplies fall as the result of an external effect (say, flood or drought), spot prices will rise and with them futures prices. If increased deliveries are expected at a later date, futures prices can be expected to fall, and so spot prices fall. In markets where delivery is permitted this ensures that the maturity date prices are the same for commodity of similar specification.

The factor which permits risk reduction is the designation of a maturity date for each contract, and the fact that on that day physicals and futures prices are identical. Trading prices will still fluctuate, and the extent to which they do will be unknown in advance and may be unpredictable. What is known, however, is that the difference between the spot and futures prices is the cost of storing the product to maturity date, and that this cost declines at a fairly regular rate until no difference exists at maturity. It is on this foundation that the risk-reducing property of the market depends.

Hedging

The process of risk reduction is known as *hedging*, which depends on taking up equal but opposite positions in the physical and the futures markets so that the two sets of price changes are offsetting. For example (below), suppose a farmer puts into store on 1 September grain valued on that day at £120/tonne, and sells a futures contract for May delivery for £140/tonne. On 1 February he has to sell his stored grain because he needs the money, although the spot price has fallen to £105/tonne. If, as might be expected, the price of futures has also fallen to £125/tonne, he buys back a contract in the futures market for the same quantity he is committed to sell (known as 'closing out his position').

Physicals/tonne	£	**Futures/tonne**	£
Sept. Stores grain	120	*Sept.* Sells grain for May delivery	140
Feb. Sells grain	105	*Feb.* Buys May delivery contract	125
LOSS	15	GAIN	15

In this situation the producer has achieved a *perfect hedge*, since the loss of £15/tonne on physicals is offset by the gain of £15/tonne on futures. His gain has been made at the expense of a speculator who bought futures in anticipation of a price rise (which underlines the fact that one man's gain on futures is always matched by an equal loss by someone else).

In this example the farmer took out a *selling hedge* against the risk of a fall in the price of products for later sale, but it is also possible to take out a *buying hedge* to guard against a rise in the price of anticipated purchases. For example, a farmer who will need to buy feedstuffs can take out a buying hedge against a price rise. A pigmeat producer unable to raise his selling price may thus guard against the risk of lost revenue/profit consequent on a price rise in his basic input.

Target-level profit pricing

Futures markets are also used to build in a given level of profit from growing a commodity (target-level profit pricing). For example, suppose that in November a farmer is asked by his merchant whether he wishes to place an order for seed potatoes for planting the following year. From past experience and current costs of fertiliser, agrochemicals, seed, etc. he can calculate production costs fairly accurately. Disappointed by his forward sale contract in the current year, he has decided not to enter into it for another year, so he cannot use that as a selling price guide. However, in November he can find a futures price for the following November and, since potatoes are a deliverable crop, if he sells on the futures market he is bound to achieve that price providing the specifications are met. More importantly, whether or not he delivers he has the option of closing out his futures contract, and if the price has fallen he will make a gain to offset the decline in the physicals market.

All he has to decide is whether or not the futures price is high enough to leave him a large enough profit once his costs are covered. In the example shown below, by hedging he has covered the cost of production and achieved a profit of £8/tonne – not quite the £10 desired, but much better than a loss of £12 if the crop were grown unhedged.

Physicals/tonne	£	**Futures/tonne**	£
November Year 1			
Cost of production	60	Sells for *Sept. Year 2* delivery	82
Profit required £10	10		
COST	70		
September Year 2			
Lifts potatoes, sells	58	Lifts hedge, buys	72
LOSS	12	GAIN	10

Speculation

The futures market can thus be used to hedge away both selling and buying price risks, and theoretically it is possible for the market to be composed exclusively of hedgers who trade with each other to spread risk. In practice this is not how the market works, and it is highly unlikely that such a market would work effectively because the volume of business would simply be too small.

In reality most traders on the market are speculators who have no physical position to protect, but use the market solely to make money by bearing risk. Speculators take a view of market prospects and back their opinion with their money. If they expect the market to rise, they will buy futures and sell when the price rises to take their profit. If they think the market will fall, they will sell futures with the intention of buying back later at a lower price, and provided their judgement is right they will again make a profit. If the price continues to climb, they will be forced to buy back at the higher price and thus incur a loss – proving the definition of a speculator as someone who is prepared to lose money in the hope of making it. The market is in other words organised so that for every gain there is a loss, and the two sums are equal: it is a zero sum game.

CONCLUSION

The price decision affects profits through its impact on revenue. The price which can be charged is a reflection of customer perceptions of the value of one product relative to competitor offers. Price is also an important marketing tool, which contributes to the creation of a differentiated product image and competitive advantage. However, it is a decision which is ultimately subject to the interplay of supply and demand in the marketplace, against which the producer of some agricultural commodities may cover himself by futures trading.

CHAPTER 9

The place decision

'Placing' the product involves all the activities necessary to make products available for purchase and to inform customers that they are available for purchase. For physical goods the primary preoccupation is invariably physical distribution (*logistics*): transportation, storage, sorting, packaging, stock and order control etc. The ability of a channel to allow effective communication between producer, distributor and consumer requires equal consideration, however, because it is the means of achieving effective customer feedback and a good match between supply and demand.

For service products like farmhouse accommodation, leisure activities or contracting services, physical distribution is obviously not involved, and communication with customers is the only way of 'distributing' the product. For service products place therefore overlaps with promotion (Chapter 11), and the term *marketing communications* is sometimes used. 'Distribution' is more generally used, and is so here, for both physical and service products.

The channels through which products pass and the condition in which they reach the consumer directly affect the revenue received. Simply placing products in the right outlets may gain a quality premium and/or reduce logistic costs; adding value may sometimes entail nothing more than the choice of another buyer or a better distribution channel. For example, improved physical distribution may allow higher-value markets to be targeted because it reduces product deterioration. In a long-run enterprise like livestock production, better communication with customers is particularly important since allows quicker feedback of market requirements and faster modification of the marketing mix. Consequently, distribution is not a job that can be seen as someone else's responsibility: it is part of the management effort to maximise the revenue from production, and the most important marketing decision which many agricultural producers have to make.

All distribution decisions interact strongly with the rest of the marketing mix, because they are about getting a product to the right place at the right time, via the right marketing channel, in the form in which it is required. Where and how a product is distributed is related to, and may even be dictated by, its form:

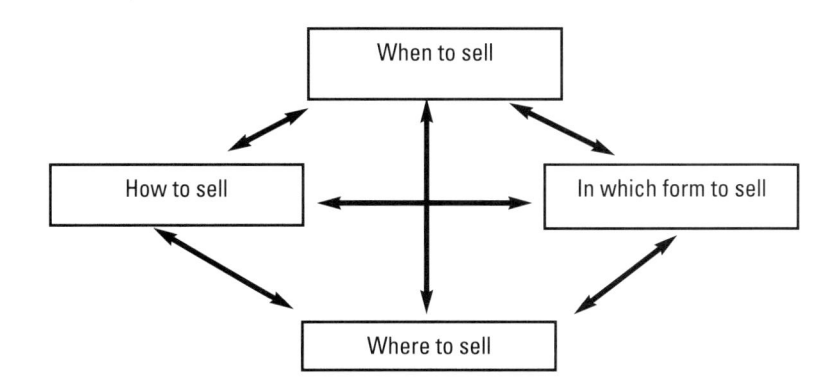

Milk and most other raw materials are collected ex-farm by the buyer/buyer's agent (where), and most are purchased only on contract which specifies terms of sale (how) and timing (when). For processed products delivery is invariably required, but may or may not be forward-contracted. Conversely, a chosen outlet may entail changes of form: a producer who wishes to sell direct to consumers will therefore have to become a producer-processor in order to provide products in the form required. Producers wishing to gain access to multiple retailers must generally use supply channels which guarantee traceability and specify place, form and time undertakings, usually on the buyer's terms.

The main temporal consideration (when) is whether products in hand will be sold immediately or stored for later sale. Grain at harvest may be sold ex-combine or dried and stored; meat and fresh produce may be frozen; even livestock may be 'stored' by delaying finishing. The motivation may be an anticipated price rise at a later peak demand period, the need to maintain consumer availability over the whole year, or to build up stocks in anticipation of a skewed consumption pattern. Processing is another means of storing perishable and seasonally skewed output (milk, butter). Seasonal pricing scales for processing inputs reflect this, plus the manufacturer's need to optimise the utilisation of installed plant (equally relevant to a farmhouse processor). Forward-contracted supply arrangements for industrial raw materials are therefore generally required as a guarantee of supply.

The place decision has strong strategic implications, because it takes time to build up good distribution channels and buyer relationships, and it may be difficult to make rapid changes in response to market and environmental conditions. Gaining access to new channels demands a much higher management commitment than using existing channels (as farmers clearly know). Where established channels exist, sales may depend on the ability to gain access, and the inability to achieve this may be a major constraint. In some countries a handful of multiple retailer buyers control access to the consumer, so the distribution decision is effectively about gaining access to these buyers.

The distribution decision must, finally, be consistent with business objectives (for example, how much marketing involvement and control is desired) and with strategic marketing objectives (is a wide geographical spread of outlets desired, or in-depth penetration over a narrow or local area?). The ability of a distribution channel to permit product branding should also be a major consideration. Will it accept producer brands as opposed to retailer own labels? If it will, will producer brands be given equal prominence with own-labels? Who in the channel is responsible for product identification, quality assurance and promotion: the producer, the retailer, or a third party? How is channel co-ordination achieved: by contractual agreement, informal relationship, or group marketing?

The answers to these questions may well determine whether a producer remains a raw material supplier or becomes a processor and/or a retailer. Many multi-product farm businesses are both, and need an understanding both of raw material and wholesale/retail distribution channels. Two chapters are therefore necessary here to cover the ground. This one considers the merits and the disadvantages of the firsthand marketing methods available to agricultural raw material producers; Chapter 10 considers the distribution of consumer products and services to retail point of sale.

MARKETING RAW MATERIALS

The primary decision for a raw material producer is whether or not to process output and/or sell direct to consumers, and it is a matter of fact that most do not, because they either do not wish or are unable to become processors and retailers. They remain suppliers of bulk, unprocessed commodities, using the network of firsthand marketing channels which have grown up to add place, time and form utility to such commodities.

Firsthand marketing channels vary substantially by product sector as well as by country and by region, because they reflect geographical and cultural divergences and the historical development of trading patterns. In the last ten years there has been a trend towards the integration of many intermediary functions and institutions, under the twin pressures of customer demand for traceability and improved channel efficiency. The sections which follow therefore focus not on the *channels*, but on the principal trading methods which are encountered in all channels: direct buyer/seller negotiation (*private treaty sales*), mediation by auction, and forward contract. The first two are spot selling methods which effectively dispose of production already to hand; the latter pre-plans production to meet a particular demand specification: it is hence more consistent with a market segmentation strategy.

Each method has its own strengths and weaknesses as well as its own procedures, which are a given. For example, forward sales cannot be made at auction; some commodities can only be produced if a contract has already been entered into. Every commodity has its predominant selling method which reflects the nature of the commodity and the buyer, product use, and history. Virtually all milk and bacon pigs produced are sold on contract, for instance; most wheat is sold by private treaty to merchants or millers; most sheep pass through the auction.

In most cases a genuine distribution choice is nevertheless possible, and it will be made with reference to business objectives (for example, price achieved, convenience) and to the level of investment and marketing commitment necessary to achieve the benefits sought. (The principal criteria for the selection are summarised in Table 9.1.) Given the long-run nature of

---

Criteria for channel selection: firsthand intermediaries

- Does the channel effectively target end-use demand?
- What is the basis of price determination: is it competitive, and if so, is the competition organised to be fair to both buyer and seller, and does it allow all parties to have a role?
- Does it ensure that the quality and quantity attributes of the product are accurately assessed and remunerated?
- What role does market information play in price determination, and is the competitive process adversely affected by an imbalance of information between buyer and seller?
- Are the terms, conditions and costs of sale known in advance, and what are the relativities between different channels/methods?
- Convenience
- Sustainability (long-run market prospects)
- Risk

Table 9.1 Criteria for channel selection: firsthand intermediaries

most agricultural production, the distribution choice needs detailed consideration *before* trading starts and decisions are made which are subsequently difficult to change. Admittedly it is possible in some cases to use a different channel next time if a chosen outlet fails to satisfy: a livestock producer may shift from deadweight sale to abattoir to the auction or vice versa; a grower may move from one wholesale market to another. This opportunity is rapidly diminishing, however, reflecting the declining number of buyers, their requirement for a continuous supply of predetermined quality, and the consumer demand for traceability.

BUYER-SELLER NEGOTIATION

In a private treaty sale the price and terms of sale are negotiated between the buyer and seller without mediation or transfer of title (product ownership) to any third party. These terms of sale must always be recorded in case of dispute: for example, in a signed sales note or invoice, which may also refer to any standard terms relating to the sale of that commodity.

Private treaty sales are the normal method of sale for a wide range of agricultural commodities, and for virtually all other goods and services produced by farm and rural businesses (farmhouse processed products sold to wholesalers/retailers; leisure & other services sold direct to consumers). The seller makes a deal with a buyer for an immediate sale at a stated price. The price of agricultural commodities may be based on a quality assessment made by the prospective purchaser, often on the basis of a sample (malting barley traded on starch content), but in many cases no quality assessment is made (feedgrains). Some livestock is bought through private treaty arrangements with dealers or slaughterers' buyers who visit farms, inspect stock and make an offer (per head, per kg liveweight or deadweight).

The principal disadvantages of private treaty trading for the seller are:

- the relative bargaining power of buyer and seller, which reflects the amount of market information which each party has, and the availability of alternative customers
- terms of trade.

Market information is a major factor determining price, and since farmers are usually infrequent sellers, their knowledge of the market is likely to be poorer than that of a buyer whose entire working life is spent trading. To redress the balance, market information about alternative buyers and prices may be obtained via telephone, media and on-line price reporting services which exist in most countries. In the UK, for instance, the MLC provides a daily price service, and the price offered by a buyer in the farmyard can be instantly compared with the average market price for livestock sold at auction that day.

This information is useful in deciding whether or not to accept a price offer, but it does not prevent producers with no alternative outlet from being offered less than the going rate, since it does not improve their bargaining position: at its simplest, one lone producer who needs to sell his product, ranged against professional buyers who do not need to buy that particular product. Nor does it remove the need to negotiate terms of sale, which are usually on a take-it-or-leave-it basis.

TERMS OF TRADE

A significant disadvantage of private treaty sales is insecure payment and long payment periods (time between product delivery and receipt of payment). Since the sale is directly negotiated between buyer and seller there is also no recourse to any intermediary if payment is delayed or not received. In the event of buyer liquidations there is a consequent risk that suppliers may incur large losses through non-receipt of payment for product delivered, which may have been processed and already sold on. As unsecured creditors, suppliers have little chance of recovering their loss – a sharp contrast with the auction market, where the auctioneer is required to keep clients' money in separate accounts to secure payment in case of liquidation.

Payment periods in the auction are also short compared with most private treaty sales: typically 7 to 14 days, compared with anything up to 100 days reported for payment from large multiple retailers. If large sums are outstanding and interest rates are high this could be very serious. The seller must therefore consider whether the scale of future sales and the size of any price advantage offset the risk of selling to firms which operate a long payment period policy, or whether a better option would be sales on 14 days' credit at a lower price.

The bargaining process is made more equal and the seller's management risk is reduced by the existence in most countries of central produce and commodity (or terminal) markets, to which producers can consign products for sale by an agent located in the market. Consignment selling via commission agents is still regarded as a significant and important marketing method in many countries, particularly for producers in remote areas. (It is also prevalent in international marketing of both commodities and processed products: see Chapter 13.) The problem in some

Another big veg trader goes into receivership

After two years of difficult trading in potatoes and field vegetables, businesses are failing at an alarming rate.

In the most recent development XYZ, one of Britain's biggest vegetable producers, went into receivership on Monday. This follows the recent receivership of PQR Marketing Organisation ... XYZ will continue to trade while the receiver assesses its financial position and tries to find buyers for part or all of it ... The firm is involved in growing, processing and distributing a range of vegetables, mainly to supermarkets, and trade rumours suggest the total deficit will run into several million ...

Among the many growers owed money are at least 17 in West Cornwall ...The receiver at PQR is also looking for a buyer. The £4 million turnover business includes ABC Potatoes, with packing and washing facilities and extensive cold stores. Growers are thought to make up a good proportion of the creditors in this case too...

Adapted from Farmers Weekly

Box 9.1

countries (notably the UK) is the decline of these markets as a result of the multiple retailers' drive for direct trading links with producers and intermediary traders.

New Covent Garden seeks fresh life

When Covent Garden moved from its 300-year-old premises in 1974 it was fully occupied ... But the operation began to shrink when it was forced to compete with [other London] markets ... In addition, many London fresh-produce wholesalers have gone into liquidation or set up distribution depots elsewhere to service supermarkets ...The Authority needs additional firms due to radical changes to distribution patterns caused by multiples buying produce direct from source, and by the shrinking independent trade.

The Grocer, 18 January 1997

Box 9.2

Professional traders on central markets charge a commission for their services, and since they make their living by buying and selling they are experts in the market – often international as well as local. Grain traders, for example, have detailed knowledge of the world grain harvest, and can advise seller-clients of the best time to sell, and whether the price offered is reasonable. Since they are paid a commission it is in their interest to use their expert knowledge and bargaining skill to obtain the highest price for their client, although it is also true that the simplest way for them to maximise their revenue is to maximise throughput rather than price.

AUCTION SELLING

Auction selling is a strong feature of livestock and some agricultural produce markets. Both are subject to criticism, and the livestock auction attracts particular disapprobation, but they survive because they have distinct attractions for the producer. The weaknesses of auction selling are well known:

- added transport and procurement costs
- quality loss involved in transportation and assembly (delay and physical damage)
- uncertainty of presentations (lots) and numbers of buyers, and resulting price uncertainty and locational variation
- product quality not necessarily rewarded
- animal welfare concern related to the livestock auction mart environment and long-distance transport of live animals.

All these objections relate to the assembly of products in a physical location, but if the product to be sold can be precisely specified by description, there is no need for a physical venue. Telephone auctions in the USA and Canada are as old as the telephone, and the electronic auction via computer network is simply an advanced form of this. Both have the advantage of overcoming the problem of remote location of producers and marts, by allowing production from remote areas to attract as many buyers as better-located production units, and thereby to compete on more equal terms. Another potential benefit is overall reduced transport costs. The spread of the electronic auction has nevertheless been slow in many countries, because producers are convinced that the face-to-face confrontation of buyers in a physical mart achieves higher prices.

PRICE DETERMINATION

The economic advantage claimed for the auction is that price is determined through competition between buyers, and not between buyers and sellers. In private treaty sales the price depends on the relative market knowledge and bargaining strengths of a professional buyer and the seller: the buyer tries to lower the price by stressing the deficiencies of a product, while the seller tries to raise it by stressing its strengths. In the auction, the competition is between professional buyers with equal knowledge of the market and more equal bargaining power. Neither the seller nor the auctioneer take any active part in price determination: the auctioneer's only official role is to record offers and the seller's to accept or reject an offer.

In practice there is some scope for salesmanship in the physical auction (though there is none in the electronic auction). In many auction markets there are long-standing relationships between buyers and sellers, and both often know that produce will pass from one to the other, and the auction's only function is to determine the price. The seller's manner and traditional cash-backs ('luck money') may have some effect on this, as will the auctioneer's skill; what many farmers gamble on, however, is that competition between buyers in a public auction, sharpened by personal rivalry and the need to procure supplies, will raise the price received.

There is no doubt that the auction process is very effective in determining the price at which supply and demand are equal, and there is little difference in this respect between the English and Dutch methods (ascending and descending bids). It is also very effective in determining a fair price for unique items like a pedigree bull. The seller does not have to price the product, and provided a sufficient number of keen, knowledgeable buyers are present, their competition will establish the price. Sufficient buyers *must* be present, however, and they must actually compete rather than collude to form a buyers' ring. The sale must therefore be well publicised, and if a physical venue is involved, the site must be convenient and attractive enough to bring buyers in. If only a few buyers are present lower prices are likely, and it is on this basis that small auctions are often criticised.

The size and location of auctions are thus important considerations, as is the quality of their management. For example, auctions with pre-registration of lots are more likely to attract buyers because they can be more certain of procuring the supplies they need (yet farmers have resisted forward notification). In the electronic auction, catalogues of supplies coming forward are always produced, now available on-line, providing detailed information about lot origin and characteristics. This substantially explains the attraction of the electronic auction for buyers, which reduces the risk of travelling to marts without any certainty of obtaining sufficient supplies.

Buyers also benefit from the reduced cost and greater convenience of remote buying from their own offices.

From the seller's point of view the electronic auction overcomes the relative disadvantage of small markets and adverse location, which independent research has shown to result in lower producer prices. It also has the potential to reduce overall transport costs, which may be reflected in prices paid, though this depends on the size of loads hauled. Traditionally, farmers have transported small numbers of animals to market, where small lots were bulked to form truck loads for onward movements. If larger loads of animals have to be collected from farms, farmers will either have to sell larger lots (with some possible loss of quality if they are not sold at optimum weight), or collection centres will have to be established and the associated costs and logistic problems solved.

QUALITY ASSESSMENT

The other advantage claimed for the physical auction mart is the ability of buyers to inspect supplies personally and make their own assessment of product quality. For products of variable quality which cannot be accurately described (for example, second-hand cars) this may be an advantage. For most commodities it presents major problems, and in some cases is unnecessary. This is particularly true of livestock, which account for the bulk of auction mart sales.

Store livestock may arguably be described as falling into the same category as second-hand cars: the basic characteristics are describable (breed or cross, male/female, liveweight, origin etc.), but there is sufficient room for variation to make inspection by buyers advisable. In the case of finished stock the situation is quite different. The buyer purchasing for retail sale buys a live animal and bids a price based on its known liveweight and declared breeding, from which he has to assess the yield in terms of various meat cuts and quality. Experimental work shows that even experienced buyers cannot achieve this consistently and reliably, so there can be no assurance that the price paid reflects the realisable quantity or quality of meat. Buyers therefore operate on an averaging basis, bidding an average price for all animals on the assumption that some will kill out better or be of higher quality: the low price paid for these will then compensate for an excessive price paid for poorer purchases.

Since no-one can accurately assess quantity or quality from the live animal, buyers must often lose by this process, and it is certain that farmers do, though neither party will admit this. Some wholesalers privately admit that they occasionally make bad misjudgements, and would prefer to buy more animals deadweight. However, the farmer's trust in the competitive advantage of auction sales combines with the strong interest of auctioneers to resist change. Even worse, their shared interest has led them to resist necessary improvements to the traditional mart which would have gone a long way to answer its critics' objections.

One major improvement which was long required (and has now been forced on the industry) was the introduction of full identification of individual animals passing through the mart, allowing traceability from farm to plate. This was resisted on grounds of operational difficulty and cost, although it was always a necessary condition for rewarding quality suppliers. As Box 9.3 suggests, it was also technically achievable long before the BSE crisis of 1996 forced it on farmers and auctioneers as a condition of trading.

The second major area for improvement – the ability to judge carcass weight and quality from the live animal by ultrasonics and image analysis – is viable only at an experimental level. Until a rapid procedure is available for in-mart use, an acceptable alternative is to bid at auction on a liveweight basis (as now) *and* on a standard quality. The buyer would then make a final adjustment to the price paid once an independent assessment of carcass weight and quality had been made. Both systems would ensure that the buyer paid for the quantity and quality of meat actually received; sellers would not only be paid for what they produced, but would receive the information on carcass quality which is necessary to identify production changes demanded by the market.

Auctioneers might be expected to resist scanning because it could represent a threat to their business (since it could obviously be carried out on the farm, with no need to assemble animals in the market). However, an electronic auction combined with on-farm quality assessment could keep them in business while simultaneously preserving the competitive situation which farmers value. Farmers and auctioneers therefore have a shared interest in opening their minds to developments without which the future of traditional auction markets may be in doubt. Many meat buyers have already withdrawn from the livestock mart because it does not meet their quality and animal welfare requirements. This has reduced the number of competitors, which must in time have an effect on the price if suppliers do not transfer to other methods of sale. Equally important, continuing opposition to live marts from the animal welfare lobby can be expected, and an electronic auction which offers the same advantages as the physical auction mart without the animal movements deserves serious consideration.

Retailers have right to know the source

Too much data, too few facts - this contradiction is emerging as the main problem facing the livestock industry as it tries to satisfy retailers' demand for traceability in meat processing. Supermarket meat-buying managers, who have complained for years about the lack of transparency through the supply chain, are being vindicated as producers at last attempt to standardise record keeping.

A positive effect of the BSE crisis ... is broadening acceptance of the principle, commonplace in other markets, that buyers have a right to know the origin of a product ... Most of the necessary information already exists. 'There are facts about cattle in computers all over the country,' says one marketing executive. 'The trouble is, these computers can't talk to each other ... yet the industry has been surprisingly computer-literate for 20 years, and the killing sector for longer.'

The Grocer, 18 May 1996

Box 9.3

TERMS OF TRADE

For the seller's point of view the auction market has a number of distinct advantages related to terms of sale, not least the local venue. The regular relationship established by trading in the same marts develops a mutual trust between sellers and auctioneers (and often buyers) which is considered lacking in other channels – necessarily so, as long as producers do not familiarise themselves with the alternatives. For livestock producers above all, it allows a dissatisfied seller to take unsold lots home, the only costs being transport and the seller's time (invariably disregarded).

Payment in 7 to 14 days, whether or not the auctioneer has been paid, is another major advantage which was noted earlier, which compares very favourably with several weeks in other channels. The auctioneer is also responsible for insurance against bad debts: a major factor influencing livestock producers' loyalty to the auction, in a sector notorious for bankruptcies and delayed payment. The costs of using the channel are also known in advance: normally a toll charged by the market owner to cover operational costs, and possibly fees for weighing animals, washing vehicles etc. The auctioneer receives a percentage commission on sales, which is an incentive to maximise the price, and commission rates are generally low. (The obligation on auctioneers to collect payment from sellers for other services may inflate the apparent cost of using the auction mart, and this has to be allowed for in comparing the relative costs of other channels, where similar costs would be incurred at a different level.)

FORWARD CONTRACT

Producer commitment to the auction rests substantially on the hope of a speculative gain resulting from competition in the mart on a given day, and the fact that personal control of a product is retained until the auctioneer's hammer descends. It is therefore not surprising that many producers are reluctant to consider contractual sale, because it entails the forward sale of a specified product to a known buyer.

Contracts exist at all levels of the marketing system: at the producer/processor interface, the processor/retailer interface, and the retailer/consumer interface. The last are generally confined to institutional markets (government, local authority, hospital authorities etc.), but they have also now developed in some product sectors where consumer buying groups are involved: chiefly organic and fresh produce. The extent of retailer/processor contracting is difficult to determine, but it certainly exists for the production of some own-label products (including farmhouse production) and may be quite extensive for other products.

Contracts normally specify the time and method of delivery and the quantity and quality of product, but they rarely specify the price. They are a normal feature of some commodity sectors: for example, highly perishable processing crops, or highly specialised products for which there are only a very few buyers and producers, where production would simply not occur without an assured (contracted) market. In other sectors contracts are a rarity because producers feel constrained by an arrangement which restricts their production and marketing freedom without any guarantee of price. Forward contracted production for a known buyer allows better forward planning, however, and reduces the risk of not finding a buyer when a product is to hand which does not match the market requirement. It also provides opportunities for coordinated value-adding activities which spot sale does not allow, both horizontally with other farmers and vertically with downstream firms.

Both buyers and sellers are nevertheless reluctant to put agreements on paper, given frequent reporting of cases where both sides have reneged on a contract.

In principle contracts are legally binding, and could therefore be enforced through the courts. In practice such enforcement is uncommon because the costs and inconvenience to either party would be too great, and this has the effect of discouraging the use of formal contracts where they are not indispensable. Alongside formal contracts the last decade has consequently seen the development of less formal supply links which are based on mutual trust and agreed business practice rather than on formal contracts (Plates 1 and 9).

Three kinds of formal contract may be distinguished which have very different levels of commitment and risk: market-specifying, production management (or transferred management), and resource-providing contracts.

MARKET-SPECIFYING CONTRACTS

In market-specifying contracts the producer retains control of all the operational production decisions and the contract assures a guaranteed outlet with a known buyer, who thus assumes part of the risk of the coordinated operations. In many contracts no price is specified in advance, so the price risk for the producer (the chief risk) is not reduced. The gain for the producer is therefore not a guaranteed, known price, but a guaranteed buyer and a reduction in the effort and expense necessary to dispose of product as it becomes available.

The buyer gains though assured supplies as and when required, which is essential in some processing situations with high fixed costs of plant whether it is utilised or not. However, the value of this is minimal to buyers of readily available agricultural products like cereals, livestock and ware potatoes, whose gain may simply be a saving in procurement and assembly costs. The disadvantage of contractual purchase to the buyer is the loss of his ability to make speculative profits when he has superior knowledge of the spot market: he is in the same position as a farmer-partner in a contract who is prevented from gaining from competition in the spot market. This has been a major factor in limiting the growth of contract marketing even when no price is specified in the contract.

In some countries there are parastatal marketing organisations (for example, marketing boards) to which producers are obliged to sell specified commodities, which the organisation is obliged to buy. These obligations are generally set down in a written contract even though the producer has no alternative except to sign. These organisations have been abolished in many countries, obliging producers to make individual contracts with processors or join marketing groups which negotiate supply contracts. The latter have very similar terms to market-specifying contracts, but the group sometimes acts as agent rather than principal, never assuming title to the product, in which case the contract is between the producer and the buyer. Where the group acts as principal the contract is between the member and the group, which is responsible for marketing, payment, administration etc. and (where it does not process) for selling on to another buyer, usually by contract.

Member-producer group contracts (usually called 'membership agreements') are legally enforceable, but in practice it is difficult to exercise effective sanctions on defaulting members because penalties are seen by some members as a contradiction of the spirit of association. Loyalty to the group rather than a legal liability therefore has to be relied on to establish an effective marketing business. In practice this depends on achieving group prices at least as high as and preferably higher than members could obtain elsewhere. For members to sign a contract binding over a substantial period there must also be a reasonable assurance that this will be achieved over the long run. Since this has been difficult to achieve in the past, members of marketing groups have tended to regard them as buyers of last resort, and adherence to buyers' contracts signed by the group has been difficult if not impossible. Where members have no alternative outlet the problem does not arise, but it is a serious obstacle to effective group marketing (Chapter 12).

PRODUCTION MANAGEMENT CONTRACTS

In production management contracts the buyer participates in production decisions, in resource specification, and in cultural/husbandry practices employed, thereby assuming a greater share of the risk and responsibility for the end product. In practice the buyer's intervention ranges from minimal to very extensive. Some simply specify the area of crop/size of output contracted for and give cultural advice (for example, sugar and blackcurrant contracts in the UK). In some vegetable production, by contrast, the buyer may specify in detail the variety, planting and harvesting dates, harvesting method, pesticide treatment etc. Many meat contracts contain similarly detailed requirements, coupled with inspections by the buyer or by a marketing group with whom a contract is signed.

The *disadvantage* of production management contracts is that the ultimate responsibility and risk of loss remain with the producer, because a crop may still fail or animals may not meet production targets, and payment is based on the quantity and quality delivered – not the amount and quality contracted for. The risk is moreover increased if the output is extremely specialised, and only one buyer exists to whom it is contracted. If that buyer reneges on the contract at the last minute the farmer is in a very weak position, which explains producer resistance to more contracted production. The *advantages* to the producer are an assured market provided the contractual conditions are met, and advice and assistance in meeting the specifications, which should reduce the risk borne. (The value of this advice has increased as other advisory inputs have become an add-on cost). The benefit to the buyer is more reliable planning, since crops and livestock are regularly inspected on farm; processing and marketing plans can therefore be constantly up-dated and revised as necessary.

The producer's principal objection to production management contracts is the fear that buyers will interfere continuously in production, but this is not borne out by experience. Many buyers regard the need to provide advice as a costly extension of their activity, into which they have only been forced by the supplier's inability *reliably* to meet marketing schedules and deliver produce consistently of the right quality and quantity. The marketing partnerships and alliances between multiple retailer buyers and producers cited in this book, which may or may not have contracts of varying degrees of formality, may therefore meet both parties' interests better provided the partnership is adhered to on both sides.

RESOURCE-PROVIDING CONTRACTS

In addition to arranging the market and assisting with production, the buyer in this case also provides resource inputs. This form of contract carries the highest producer risk *not* because of the contract itself, but from the indebtedness which results from accepting inputs.

Resource-providing contracts nearly always provide working capital on which the return is high over a short period, leaving the producer to find his own fixed long-term capital which has a low return. This has sometimes led farmers to use any small collateral they have to back fixed capital projects (for example, chicken and pig housing) with which to produce output from resources loaned by buyers at high interest rates, when a better use would invariably have been to back working capital to generate more income from existing enterprises.

A guarantee of purchase is generally given, but no guarantee of the price to be received for the output. The non-farming partner provides all the required inputs and invoices them to the farmer for payment when the product is supplied. When delivery is made outstanding amounts are deducted from the sales revenue due to the producer, and the balance is paid to him as recompense for the use of his fixed resources (land, buildings etc.) plus his management and labour. The danger is that if the market price has fallen during the contract, the producer may owe money to the buyer which the latter will not unreasonably seek to clear. Producers are also generally required to purchase inputs from the buyer, which prevents them from shopping around for lower price inputs (and there is, of course, no opportunity to shop around for a higher product sale price). By contrast, both of these options would be preserved by taking out a structured loan with a commercial bank.

The problem in this situation is not the contract as such, but the capital borrowing and the mismatch between the short length of the contract and the need for long-term investment to comply with its terms. There are cases where such contracts are perfectly satisfactory. For example, where a processor wishes to encourage production of a new product which entails a high production risk, he should reasonably expect to bear some of that risk because he, as the innovator, is likely to make high profits. The priority for both partners is to ensure that producers taking up the contract have the necessary skills and track record in similar crops or enterprises, to improve the odds of achieving a high level of success.

TERMS OF CONTRACTS

For buyers seeking to obtain the exact product specification they require, contractual production may be the ideal (some would argue the only) marketing method. The advantage to the *supplier* must always be assessed by reference to his objectives, enterprise mix, pattern of production, and the precise contract terms. It is therefore not surprising that standard buyer contracts are accepted by some producers and not by others, and this should not be interpreted as meaning that the contract is unreasonable or unfair: simply that it does not suit every business.

In negotiating or evaluating a contract several points must be settled to the mutual satisfaction of the contracting parties:

- price
- quality required and its assessment
- quantity required
- terms of trade (date of delivery and payment, weighing conventions, etc.)
- fairness and sustainability.

PRICE

Price is often completely absent from contracts, or is specified only as 'the price ruling at time of delivery'. This represents a risk which the producer has to evaluate relative to other conditions and potential benefits. A fixed price contract is not the solution, however, for if the contract is entered into before production starts it is impossible to predict output accurately, and hence what a reasonable price will be. If there is a free market for the output and the price in the market exceeds a contract price, producers will simply renege on their contract and sell in the market to make a speculative gain. If the market price is below the contract price the buyer will renege or raise his acceptance standards and buy in the free market to make a similar speculative gain. Various forms of formula pricing have been devised to prevent both situations arising.

Basic formula pricing

In basic formula pricing, contracting parties agree to use a specific published price as the basis for their own calculations, together with agreed adjustments to this base price, for the duration of the contract. These adjustments allow for quality differences, delivery time, and even eventual selling price. The price selected as the base is critical, since it should be representative of ruling prices for the same commodity. In many countries a published series of prices is accepted as the base: in the UK, for instance, the EU target price may be used, or for livestock, published weekly auction market prices. Futures market prices may also be used for certain commodities. The problem is that the prices may become self-perpetuating: the price determined by one formula may become the base price for someone else's formula, so a levelling of prices occurs which may not accurately reflect the competitive situation.

Other factors which need careful consideration are:

- the extent to which the base price may be manipulated by either party
- whether the base price provides sufficient flexibility to reflect regional and local differences
- the applicability of the reported prices to the commodity/product which is the subject of the contract (for example, carcass or retail meat prices and live cattle sales).

System pricing

System pricing overtly recognises that the critical price is the one paid by the final consumer (whatever other members of the marketing channel liketo think); this therefore becomes the basis of the pricing system. The method arrives at a price for each successive business upstream of the consumer by a process of cost deduction. Ideally this would be the consumer price, but in prevailing market conditions the 'consumer price' is that which a retailer is prepared to pay. Once this base price has been agreed, the total production and processing costs are deducted and the profit is divided between the two parties according to a negotiated formula.

For example:

	p/kg
Beef carcass meat	225
Production costs (farmer)	145
Slaughter/wholesale cost	45
Profit	35

This is effectively a fully integrated business agreement, with full knowledge of each party's business being available to the other, and it has the capacity to develop greater efficiency in both production and processing. This formula also has the advantage of flexibility, since the base price will move as a result of changes in supply and demand and will influence the price received by processors and producers. Similarly, cost changes will be reflected in received prices. As with all formula pricing methods, speculative gains are difficult to achieve, but all parties benefit from increased prices and reduced costs and all lose in the reverse situation. A modification of this method is *activity-based costing* (cost-plus pricing as defined in Chapter 8), which adds up the costs incurred by all parties to arrive at the final sale price.

Pricing by tender

Formula pricing depends on the existence of a market which is believed to represent the true supply and demand situation; *system* pricing will work only where contracting parties are prepared to share information fully. Where neither condition holds, the problem is how the price can be determined, and the simple answer is *pricing by tender*. The buyer would publish his contract terms, and suppliers would submit offers of the price they would need in order to meet these terms and sign the contract. This would be a fair and competitive system, allowing producers accurately to assess their total costs and to make offers on this basis. The only debatable point is whether each supplier should be paid the price he tendered, or all suppliers be paid the same price, which would have to be the highest in order to achieve sufficient supplies. (In Figure 9.1, should you buy Q1 from seller A at P1, Q2 from seller B at P2 etc., *or* pay P3 to sellers A, B and C?) In economic terms, can the buyer price-discriminate against producers, and therefore minimise his own costs?

The question is one of fairness, but the difficulty is to decide what is actually fair. One argument is that if the producer has calculated his costs correctly and offers to supply at a lower price, it would be fair that he should receive the price he tendered: if he makes a mistake, that is his problem. Farmers in the past have argued, conversely, that it is unfair for two farmers to receive different prices for the same article (although they do not argue that in the auction). In some cases they have lobbied governments successfully to ensure that this does not happen. In other cases they have informed each other of offer prices or formed a bargaining cooperative to ensure that everyone receives the same price, and price discrimination by the buyer is prevented.

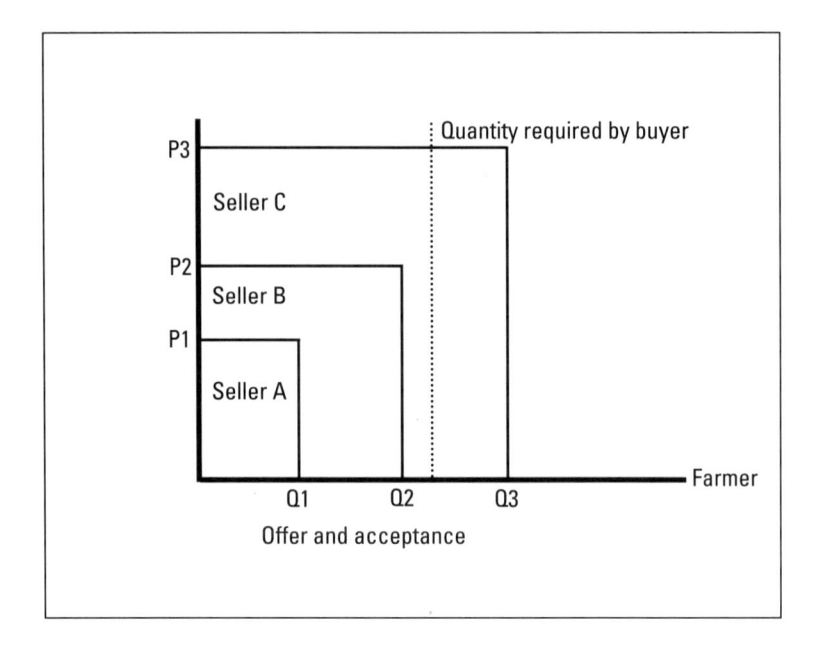

Figure 9.1 Pricing by tender

QUALITY ASSESSMENT AND REMUNERATION

Quality specifications are crucially important in contracts, but it is sometimes difficult to define quality in terms which producers can understand and, more importantly, can adjust production to deliver. For example, many potato contracts specify a minimum size, absence of green colour, cracks etc., and some vary the price according to various percentages of size groups, but in field conditions these apparently simple qualifying conditions are difficult to meet. Similarly, the nitrogen content of malting barley or the Hagberg number for wheat may be difficult to achieve. In the case of milk a high degree of technical precision exists, with objectively measurable parameters (fat, protein, lactose, total bacterial count, somatic cell count, absence of antibiotics and added water), and supply contracts vary the prices accordingly.

Many contracts for processing inputs include an escape clause which refers to 'product unsuitable for processing', which gives the buyer immense discretion in deciding the acceptability of the product. For most products it should be possible clearly to specify objectively measurable, required technical quality standards, but if this is not so, and escape clauses are included, the contract should also include defined arbitration procedures.

QUANTITY

The quantity of product contracted for needs to be specified in the contract and therefore needs to be negotiated. The simplest solution is to contract for the total farm production of a given commodity (for instance all milk produced), or total production of a given type or variety (all bread-making wheat or plum tomatoes, excluding all other wheat and tomatoes). This leaves the buyer to bear the risk of any excess or deficit of product to maintain his production. It may also allow fraudulent product diversions by producers when product is scarce and prices high.

The alternative is to specify a given quantity (the 'absolute quantity' method), possibly with a small acceptable variance. This enables the buyer to guard against massive oversupply (though it does not protect against failure to supply) and to purchase the quantity he requires, leaving the producer free to sell any excess wherever and for whatever price he can obtain. In practice both methods are used. For perishable, short-life products total production is more common, but the absolute quantity method is easy to operate and is used widely for storable commodities.

TERMS OF TRADE

All satisfactory contracts will include details of how they will be implemented, which need careful scrutiny by the potential supplier. Payment arrangements are the most obvious detail: for example, the 15th of the month following delivery – a form very common in fresh product contracts. Where delivery is made immediately after harvest and subsequently stored for sale up to a year later (wool, vegetables used for freezing) stage payments may be specified: for instance, 30 per cent seven days after delivery, 30 per cent sixty days after, 40 per cent 150 days after. Where farmer-controlled businesses are the buyer this is usually accepted, albeit reluctantly; in other cases it may simply be imposed by the buyer as a condition of trading. In evaluating the trading opportunity the supplier therefore has to calculate the costs to his business of delayed payment and compare this with the benefits of producing that commodity, with reference both to increased profitability of the commodity itself and any cost-reducing/revenue-enhancing effects on other enterprises.

Security of payment is also critical to most businesses but presents a particular problem in assessing contracts. However good the contract, a firm which has already gone bankrupt and ceased to trade cannot pay for product already acquired. A supplier, who is unlikely to be a preferential creditor, will therefore have to wait for the receiver or liquidator to resolve the financial affairs before any payment will be made: a marked difference from the auction, where sales are often covered by indemnity insurance which pays suppliers in cases of business insolvency. A buyer's track record may be all a supplier has to rely on in evaluating this risk, but the buyer's credit rating can be investigated and an indemnity policy taken out to protect against loss.

Further details which must be built into the assessment include transport arrangements and costs, and arbitration arrangements in cases of dispute about quality and quantity etc. For example, assured regular collection is imperative for a highly perishable product like milk, for if it is not collected the producer not only goes unpaid, but there is no storage for the next milking because vats are full.

FAIRNESS AND SUSTAINABILITY

Contracts may have a limited, specified duration (one year, one growing season) or they may continue for as long as the parties wish them to remain in force ('evergreen'). In the latter case the contract must also incorporate termination conditions.

Most businesses entering a contractual relationship normally expect – or at least hope – that it will be sustainable over the long term, especially if they are tailoring their production to meet a buyer's product specification. Both parties must therefore be able to agree at the outset that the contract is fair and equitable, or it is unlikely to be sustainable. Contracts of themselves do not ensure fairness; the terms may do so, but far more important to a working relationship is the atmosphere in which the contract is negotiated. Only if the parties accept that there is a mutuality of interest in the terms themselves, and in the way in which they carry out their obligations, is the trading relationship likely to be sustainable. If the parties seek to work the contract to their own advantage or are not wholly committed to the undertaking, the relationship will break down, however well the contract is drafted and however fair its terms.

Outcomes satisfactory to both parties consequently depend on trust rather than on paper agreements, and trust is only built up over time, on a track record of reliability and fair dealing. Contract appraisal should therefore concentrate as much on track record and reputation as on legally enforceable contract terms, which will only be needed if something goes wrong. In the past farmers have been wary of contractual production because they feared that vertical coordination of production, processing and distribution threatened their management independence. In today's market conditions there must be greater open-mindedness to the opportunities for marketing relationships and contractual trading, if consumer satisfaction and business profitability are to be achieved.

CONCLUSION

In all commodity markets there is a movement towards vertically coordinated, pre-planned marketing. This has always characterised some sectors (milk, some specialist processing commodities, pigmeat); it is now true of commodities like potatoes, horticultural crops and, finally, grain, beef and sheep. The driving force has been retailer pressure for complete traceability and assured levels of quality and supply at consistent prices, and for improved cost and logistic efficiency throughout the system. Farmers cannot realistically resist this pressure, nor should they wish to, since vertical coordination delivers long-term security for producers as well as buyers, and rewards quality suppliers. The challenge for intermediary traders is to modify and improve their activities to deliver the same benefits.

Budgens is currently promoting its West Country beef with the aid of a special British Meat leaflet produced in conjunction with the Meat and Livestock Commission.

John Fawdrey, senior buyer for the chain, told The Grocer that he obtains all his fresh beef from selected rearers in Devon and Cornwall. This is done with local aid from both Jaspers' slaughterhouse and Kerry Foods' processing plant that dissects the carcases into primal cuts.

Work on the initiative started in the spring in preparation for a summer launch. New labels have been designed and the packaging is carried out at the company's fresh meat depot at Wellingborough which is supplying meat counters in 66 out of Budgens 102 stores.

Fawdrey specifies breeds, weights and age in order to obtain the quality required by the multiple. Only meat from steers between 20 and 26 months is purchased and they must be South Devon, Limousin or Charolais crosses reared in the traditional manner – out at grass in the summer months and fed natural forage and cereal based rations indoors in the winter.

"We are getting consistency now," said Fawdrey. "After the last scare in March I went for extra quality. We have constant checks all along the line and the maturation period has been extended to a minimum of 14 days to improve the tenderness and succulence. We very rarely get a customer complaint on eating quality."

Fawdrey emphasised that technologists and buyers representing Budgens also make regular visits to the West Country farms, abattoir and meat packing plants to check the animals' welfare and to ensure the safety and quality of the meat.

Promoting a quality West Country image for fresh beef

Plate 1. Producer-retailer marketing alliance

Plate 2. What marketing means: the competition

Plate 3. High-welfare marketing

SAINSBURY'S

WEST COUNTRY FARM REARED
VEAL

Specially selected from farms with the highest standards of animal welfare

PRODUCED UNDER THE SAINSBURY'S PARTNERSHIP IN LIVESTOCK SCHEME

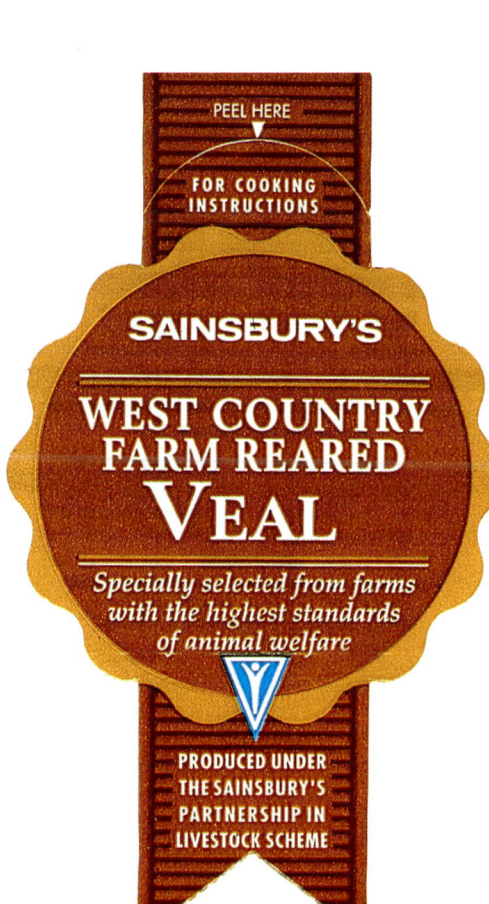

PEEL HERE ▼ FOR COOKING INSTRUCTIONS

Plate 4a (above) and 4b (right). On-pack labelling: adding value through information

PEEL HERE ▼ COOKING INSTRUCTIONS

BONE IN SIRLOIN CHOP
Remove all packaging. Grill under high heat for 7 minutes per side. Always ensure product is piping hot before serving.

VEAL SHOULDER CHOP
Remove all packaging. Grill under high heat for 6 minutes per side. Always ensure product is piping hot before serving.

VEAL SHOULDER JOINT
Remove all packaging. Place in a pre-heated oven, 108°C; 350°F, gas mark 4. For 40 minutes per 1/2 kg (1lb) plus 40 minutes. Always ensure product is piping hot before serving.

Sainsbury's Partnership in Livestock Scheme is an alliance between Sainsbury's and our British farmers and processors, which has been in existence for a number of years.

We work together to develop and improve animal welfare standards and the quality of our meat. An integral part of the scheme is to ensure animals:-

1. Reared in large straw bedded barns.
2. Have constant access to good supplies of food and water.
3. Drugs are not given other than for medicinal purposes and then under veterinary supervision.
4. All come from farms where good farming practices and stockmanship ensure animals remain healthy.

Plate 5. Adding value by packaging and product enhancement

Ready meals set to stir the market

Birds Eye Wall's has added a further 13 products to its ready meals line-up.

For the stir fry market it has added three new recipes under the Simply Stir label – chicken and prawn paella, potato with traditional mince, and chicken with pasta Provençale. They have a recommended retail price of £1.49 for 340g.

Two pub grub-style recipes based on established Birds Eye products are curry & chips, and chilli & chips 300g for £1.39).

To the lasagne sector comes a new twin pack at £2.79p for 660g, while sliced roast meats include pork and lamb varieties (8oz or £1.79).

Meeting demand for products at the 99p price point come macaroni cheese, macaroni cheese and ham, chilli con carne, and vegetable curry. Packs are 285g.

Ready meals are predicted to reach £575m this year, representing a 5% increase on 1995, making it the largest frozen food sector. The market comprises three distinct sectors. International meals, which account for 51% of value, have been fuelling the market's growth. They saw a rise of 13% year on year. Traditional dishes at 36% rose only 0.5%, while healthy lines, which account for a 13% value share, declined 5%. Sales director Tony Pearce said: "Ready meals represent a growth opportunity which we are committed to developing. We have devised a dual strategy to ensure long term success. We are protecting and building the high cash market where we are strong with Panflair, while introducing new formats and recipes at the critical 99p entry point.

"We are doing this in both areas with products which deliver quality, choice and value for money."

Plate 6. Inter-product competition

Plate 7a. Rachel's Dairy before: a farmhouse operation.

Plate 7b. Rachel's Dairy after: an industrial unit employing 30 full-time staff in processing and management.

Plate 8. Electronic point of sale scanning brings instant stock control and customer information

Plate 9. Retailer-producer alliances bring product and supply assurance and a secure market outlet.

SAINSBURY'S FARM ASSURED BEEF

Sainsbury's farm assurance scheme, Partnership in Livestock, was established in 1990. It is a vital link between Sainsbury's, the processor and the farmers. The link allows our customers to be certain that farmers supplying Sainsbury's are committed to good stockmanship and sound animal welfare, ensuring that the beef you buy from Sainsbury's is of the highest quality and from the best farms.

farm assured

WHEREVER YOU SEE THE SAINSBURY'S FARM ASSURANCE LABEL YOU CAN BE SURE THE BEEF IS OF THE HIGHEST QUALITY

ALL SAINSBURY'S BEEF COMES FROM FARM ASSURED HERDS

SOME FARMERS WILL DELIVER ONLY 5 CATTLE A YEAR, WHEREAS OTHERS MAY DELIVER OVER 1,000

WHAT DOES "FARM ASSURED" REALLY MEAN?

One of the main aims of the partnership is to provide Farm Assured Beef to all Sainsbury's stores.

The principles are simple and based on a solid belief in good animal husbandry: all suppliers are required to work to the Sainsbury's Policy on Management of Livestock Welfare and Veterinary Medicine Usage. Welfare standards are verified through independent audit, as well as visits from the processors' and Sainsbury's technical teams.

All cattle are reared and managed with care in a humane manner.

- Balanced diet, with open pasture grazing during summer months and wholesome conserved forage and concentrates whilst housed.
- Access to fresh, clean water.
- Natural light when housed.
- Good standards of housing for the winter months.
- Appropriate handling facilities.

The Partnership in Livestock deals with a wide variety of committed farmers. All farmers who achieve the required standards are welcome, regardless of the size of their herds.

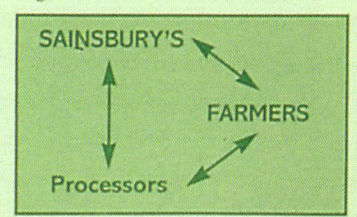

```
        SAINSBURY'S
         ↑    ↘
         ↓      ↘
                 FARMERS
                ↗
        Processors
```

The close and regular contact between the groups means that there is a full understanding of the welfare requirements, and can ensure that the standards are being met at each stage of the chain.

Plate 10. Generic promotion of an undifferentiated raw material

Plate 11. Product branding and assurance underpin a long-run marketing programme

Bramleys now benefit from a brighter image

Ian Mitchell: 'Multiples were even thinking of delisting the variety'

English apple growers are looking forward to a good Bramley season, strengthened by the knowledge that not only is there a lighter crop, but their fruit appears to have shaken off its old-fashioned image.

Ian Mitchell, chairman of the Bramley campaign group, told The Grocer: "In the early nineties major multiples were not enamoured with Bramleys. There were rumours that it might be delisted to make room for more dessert varieties."

Bramleys have also been under pressure from convenience foods and changing eating habits. Occasionally, they faced the added ignominy of having to compete against varieties like Granny Smith recommended as cookers.

"Bramley was well known and liked by certain customers but its brand image was weak," admits Mitchell.

The variety is, in fact, big business. This year's crop, estimated to be around 118,000 tonnes, has a retail value of £90 million.

Mitchell puts the change of trade and consumer attitude down to the industry's own efforts to get the fruit better known. Four years ago its 500 or so growers launched a voluntary promotional campaign to stress the apple was not just a quality product but under-used.

The results have been impressive. For example, last season a Bramley week led to volume increases of up to 80% by major retailers including Sainsbury, Tesco, Safeway, ASDA, Somerfield and Waitrose. Overall fresh sales were up 35%.

Peter Fry, Tesco's produce trading director, went on record as saying he was baffled by the scale of the growth, as it bucked the trend for convenience food.

While 65% of Bramleys are sold fresh, the remainder goes to processors. This sector showed a 25% boost and variety was named on more apple products.

For 1997 Mitchell has plans for a similar package of events which will be launched next month. They include a Bramley week in February, Bramleys for Breakfast in May, and an apple pie competition in Nottinghamshire where the variety was discovered. Safeway is participating.

Plate 12. Using a distinctive product label helps to achieve brand recognition.

Plate 13. Corporate imaging.

Plate 14. In-store merchandising

Plate 15. Enhancing the psychological content of a product

Plate 16. New meat cuts for variety and added convenience

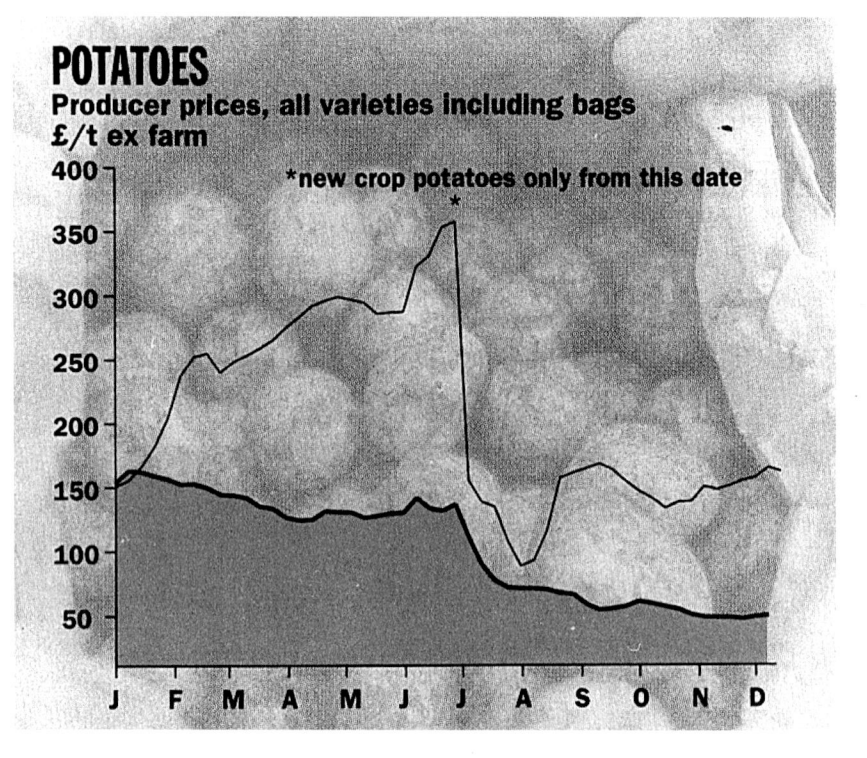

Plate 17a. Published seasonal price of potatoes

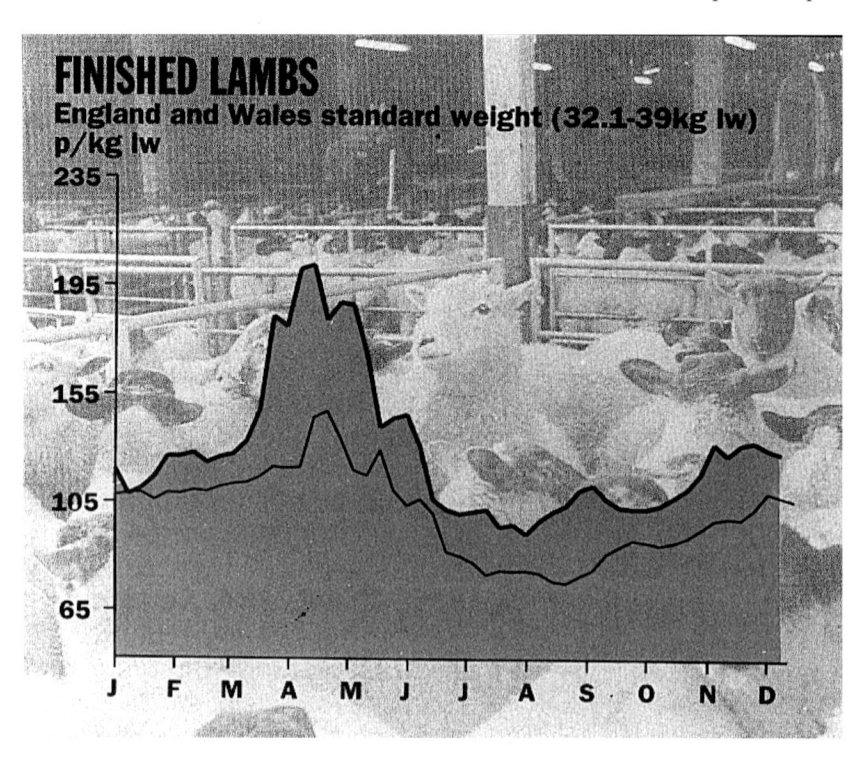

Plate 17b. Published seasonal price of finished lambs

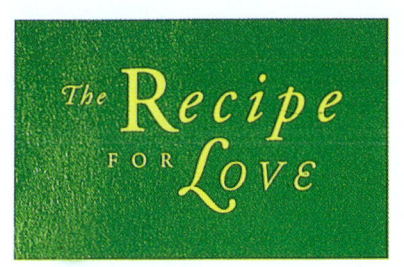

British Lamb is on TV throughout Great Britain through September and into October.

Two popular The Recipe for Love commercials will be seen by over 80% of your customers. In fact, virtually everyone who watches ITV, C4, GMTV and Satellite TV.

Tasty, easy-to-cook lamb is increasingly in demand, so sales are really set to grow. For more information and advice, call your Area Account Executive or the Retail Promotional Team on 01908 677577.

Plate 18. Trade promotion to coincide with consumer advertising

Plate 19. Producer group product branding

Plate 20. Small farm producer wins top export award

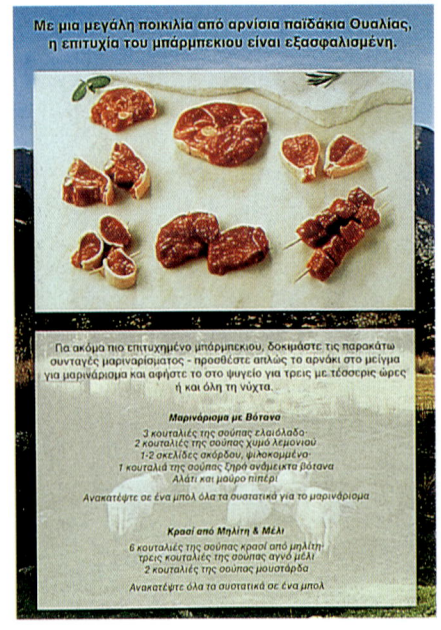

Plate 21. Producer group export promotional material

CHAPTER 10

Selling the product

Although selling is not synonymous with marketing, as we began by saying, retail sale is crucially important because the retailer controls access to the consumer via shelf space, and collects the revenue earned for production and marketing via the checkout. For producers of finished products (milk and dairy products, fresh produce, handcrafts etc) this may suggest that direct selling to the consumer must be preferable to indirect distribution via independent wholesale and retail distributors. Producers of service products have no choice, since services by their nature are supplied direct to the consumer (which may be a strong enough disincentive to rule out an otherwise viable enterprise).

An informed judgement between direct and indirect selling can only be made in the light of a thorough understanding of the role and functions of a distributor, since either the job will have to be done by a direct seller, or an indirect seller will judge potential distributors by how well they perform these functions. As Chapter 5 explained, the role of the distribution sector and the functions it performs continually evolve in the search for competitive advantage and improved efficiency, and in response to technological change. The core activities nevertheless remain the same, and were outlined in Table 5.1.

None of these is beyond the technical capability of most producers, as many successful producer-retailers have demonstrated. However, as Figure 10.1 illustrates, the producer-retailer is single-handedly responsible for all the functions which, in a conventional marketing channel, are undertaken by a number of intermediaries. What may therefore be in doubt is the ability of the owner-manager to do all this without risk to the rest of the business and to total revenue. The ability to do the job as *effectively* and as *cost*-effectively as specialist intermediaries may also be in doubt.

What motivates many would-be producer-retailers is the hope that they will achieve the middleman's share of consumer expenditure, but this hope is rarely realised and is never a sufficient reason to venture into direct selling. Cutting out the middleman transfers his costs to the producer, so any potential increase in revenue must be set against increased costs. Direct selling can of course contribute to lower cost levels, but at the expense of a significantly increased management commitment which may adversely affect the owner-managed business (possible lost revenue resulting from product quality reductions, missed deadlines, timeliness of farming operations etc.).

The sales skills required to generate and sustain retail sales are also commonly underestimated. These can of course be learned, or sales personnel may be employed, but there may be no time to acquire the necessary skills or to supervise employed staff, and there is a high cost involved in delivering a competitive offer. There is also a risk of achieving only a narrow market penetration which limits the growth potential, and may only be overcome by major investment and business development on an unexpected and unwanted scale.

There is consequently a persuasive argument for *indirect* selling, since it allows a producer to transfer selling responsibility to specialist distributors and concentrate instead on what he does best: production. The risk of lost revenue resulting from conflicting management demands is avoided, and lower investment is required (in terms of facilities and labour, including own time). By

Conventional Channel

Producer
- Design
- Produce
- Brand
- Price
- Promote
- Sell

Wholesaler
- Buy
- Stock
- Promote
- Display
- Sell
- Deliver
- Finance

Retailer
- Buy
- Stock
- Promote
- Display
- Sell
- Deliver
- Finance

Consumer

Vertically Co-ordinated Channel

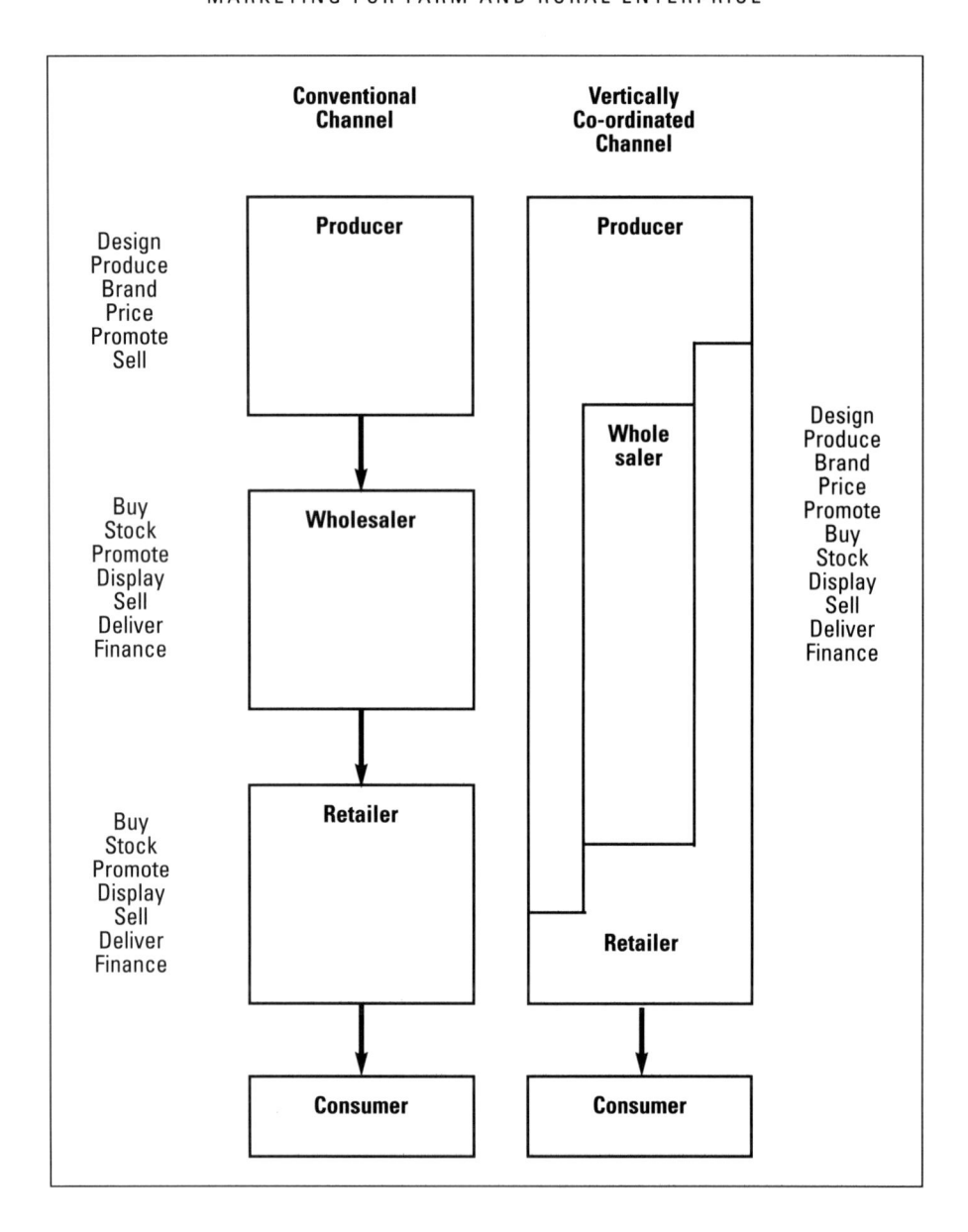

Producer → Wholesaler → Retailer → Consumer
- Design
- Produce
- Brand
- Price
- Promote
- Buy
- Stock
- Display
- Sell
- Deliver
- Finance

Figure 10.1 Distributor channels: conventional and vertically coordinated

using established marketing channels market access is guaranteed, which may be a particularly important consideration where consumers are wedded to particular supply systems (a favourite or local retailer multiple, for instance). Access is generally achieved over a wider geographical area, and specialist distributors should achieve greater distribution efficiency and lower costs than the small producer/entrepreneur marketing direct. On the negative side, downstream intermediaries who handle output from many suppliers may not 'push' a product, since it is simply one component in their product range/marketing mix.

For most businesses the ideal is therefore to identify a distributor (or a distribution channel composed of a number of firms) who has shared business objectives and values, with whom a long-term marketing relationship can be established. This achieves the main advantages of direct selling (control of distribution + effective customer feedback ➔ faster and better modification of the marketing mix) without the costs and risks which it involves.

For the small business this is much easier said than done, and a producer may simply be obliged to use any channel which will accept his product, or become a retailer himself. At the other extreme, a producer with a unique product for which consumers are clamouring may be in a position to choose whom to supply. This may seem entirely desirable, but the need to allocate supply to selected distributors needs careful handling to ensure that rejected distributors are not alienated, and distribution channels are kept open which will be necessary when supply increases and the clamour has abated.

The reality for most producers again falls between these two extremes, and the selection process usually starts with a list of preferred distributors which reflects the producer's own priorities, and only when these fail to show interest is the net widened to seek out others. The stages and the decisions involved are shown in Table 10.1, which provides the framework for the following discussion.

DISTRIBUTOR SELECTION

The physical and economic functions performed by distributor channels vary widely, and substantial differences of emphasis exist in the services provided at wholesale and retail level, as well as by different wholesalers and retailers. Chapter 5 noted the dominant channel position of multiple retailers in the agri-food sector of many countries and the associated decline of the wholesale distributor sector; wholesalers nevertheless remain an important supplier to the small independent food retailer everywhere, and in most other non-food product sectors. The choice between wholesaler and retailer will thus tend to be very country-specific, depending on the marketing channels which exist and the relative importance of wholesale and retail sectors in different countries; it will also be product- and market-specific.

The selection of a distributor will normally begin with a list of retail outlets generated by the market research, which will have identified the outlets used by targeted consumers: up-market multiple retailers, small independent retailers, discounters, 24-hour convenience stores etc. Sometimes a channel may not be obvious (most obviously in export and new product markets), so 'prospecting' may be necessary to establish whether one exists, and if it exists, whether it is accessible. The more unusual or innovative the product, the less likely it is that a distribution channel exists for it. In this case distribution may have to be sought through systems which handle related products or a product range consistent with the new product, or systems which sell unrelated products to the target market. Where neither is possible it is necessary to create a new channel – an option to be avoided if at all possible by the small business.

Sometimes a channel exists which is not accessible by a newcomer. The law may limit the selection to officially approved distributors, or restrict access by restricting the range of goods which may be sold in the same premises. (Restrictions on food products alongside other products are very common, but vary from country to country.) Access may also be prevented because a channel is tied to existing suppliers. Small brewers in the UK face this problem when seeking sales in outlets owned by large brewers whose concern is naturally to sell their own product. Many franchised outlets are tied to a supplier who is often the franchise owner, the object being to ensure distribution for his own products. Some distribution systems are composed of wholesaler and retailer partners (including caterers) to which access by new suppliers is difficult. Some are vertically integrated systems in which a supply agreement made between a producer and a distributor is imposed throughout the entire system, either through voluntary cooperation (as in voluntary chains like SPAR) or ownership (as in retail multiples and catering chains).

The intensity of coverage sought (the number of outlets served and the extent of geographical cover) is a major consideration in identifying possible distributors, and will normally be limited by production capacity, logistics, and objectives. However, it is common for a small supplier faced with a high demand for an innovative product to be carried away by success, and wish to expand production sharply to expand coverage. The risk is that as availability rises and the novelty of the product declines, demand may not increase sufficiently to make the additional investment worthwhile. The cost of a small increase in distribution intensity may be substantial, and will include costs involved in maintaining sufficient stocks (inventory) to meet demand as well as transport charges. A balance must therefore be found between the costs and benefits of increased coverage, which is the point where the extra costs = the extra revenue gained by supplying more outlets. It is also invariably better to be certain of supplying some outlets well rather than sacrifice quality of service to expansion objectives.

IDENTIFY ALTERNATIVES

- Target market
- Intensity of coverage
- Additional features

SELECT TYPE OF CHANNEL

- Availability
- Legal considerations
- Control/supervision
- Revenue, cost, sales

SELECT ACTUAL FIRM

- Independent, multiple
- Product range
- Selling method, style
- Terms of trade expected
- Reputation

GAINING ACCESS

- Product presentation, sales promotion
- Negotiation

DISTRIBUTION MANAGEMENT

- Logistics
- Follow-up support, administration
- Evaluation
- Replacement

Table 10.1 Distributor selection

REFINING THE SELECTION

The type of channel selected will reflect internal business factors and judgements about potential distributors:

Internal factors	*Distributor factors*
• Scale of production	• Sales generation capacity
• Nature and status of the product	• Independent or multiple chain
• Management implications	• Product range
• Control required/desired	• Selling method
• Cost	• Terms of trade and reputation

For example, if the product is a basic one produced in large quantities and sold to most consumers (milk, bread) it will require distribution to all types of food store. This will entail direct transfer to multiple retailers and some independents, and indirect transfer via wholesalers to independent grocers and other food outlets. If it is an expensive hand-crafted product, it may call for distribution via specialist wholesalers or direct to a handful of top retailers. Some very small producers of very high quality food and handcraft products have successfully targeted their limited output at one or two top retail outlets in a capital city, securing all the sales they are able or wish to make at a premium price.

Where products require controlled conditions to prevent deterioration or damage, outlets are required which guarantee this throughout the distribution chain (transportation, warehousing, retail premises). For food products these conditions are tightly regulated by law, so the absence of appropriate facilities at any point in the chain (or lack of confidence in their effective management) will eliminate some distributor options.

Other products may only be saleable if services are provided with them, which means that distributors must be identified who are able and willing to provide the service. The service may seem relatively minor to the producer: for example, in-store display and demonstration, consumer advice on use of a novel product. Its provision is a cost to the distributor, however, and this cost will be measured against the likely return on space/time relative to other products stocked. If the provision of a specialist service or a high level of service is necessary (for example, hand slicing and wrapping of a premium delicatessen product to enhance its prestige image, or retail presentation of meat), training may have to be provided for the distributor and all the employees who become involved over the duration of the trading link. Constant monitoring will also generally be required to ensure adequate provision. The attitude which customers encounter from the distributor during the provision of this service will also substantially influence their satisfaction and possible re-purchase. Consequently, it is not just the ability to provide the service which is in question, but the way in which the sales opportunity is handled.

Where a high degree of risk is involved in this transfer of responsibility the producer will generally seek to shorten the distribution channel: ideally producer → retailer. This is particularly important for fresh products which require effective chilled storage, stock rotation, and effective removal of out-of-date products. It may also be necessary for innovative products which require explanation to the consumer (including innovative food products). The resulting need for control may conflict with the supplier's desire/need for widespread availability, but a similar degree of control may be achievable through the use of specialised wholesalers, and through wholesaler-retailer partnerships which share the same high standards. The post-BSE recovery of the UK beef market in 1996, for example, was substantially due to producer-processor-wholesaler-retailer links which guaranteed traceability, and the delivery of specialist products with an extremely high production and marketing specification.

The cost of using any distributor must be estimated in relation to its likely sales generation capacity. If substantial sales growth is required (for financial viability, for example) a distribution system able and willing to adjust to expanding volume is essential. Given that logistic costs are commonly in excess of 20 per cent of retail price, it is also essential to choose a channel which keeps these costs under control while providing a high level of service. A distribution system with a wide depot network serving independent shops may achieve high levels of consumer availability at low cost, but so too may a multiple retailer with a dedicated centralised warehouse and store delivery system. Both independent and multiple channels therefore need to be evaluated.

Independents/multiples

At wholesale and retail levels there are small independent distributors with one or a few depots/shops, alongside large businesses with tens or hundreds of units. Small-volume owner-managers generally find it easier to deal with independent entrepreneurs like themselves, so they typically start by distributing products through local independents, establishing the initial links and subsequently supervising them by personal visit (self or employed sales personnel). As businesses grow, or if they start at a sufficiently large scale, wider distribution is required to achieve a viable sales volume. The producer who wishes to continue to sell to independents can achieve this by seeking a local branch listing in the regional depot network operated by national wholesalers – particularly the growing number of these who seek locally-sourced and regional specialities to supplement their standard offer (a strategy which also maximises the producer's local reputation and advertising).

In principle, similar access may be gained to the multiple retailers who operate regional distribution depots for their stores. The main problem is likely to be insufficient volume of supply to meet the demand of a large national chain, but in most countries these chains are stronger in some regions than others, and there are regional multiples which only exist in one area. It may therefore be possible to place products in one or two regions only, which substantially reduces the required scale both of production and logistics. If and when production expands, distribution can then be extended from this base area and may eventually cover an entire country, and even export markets in which the firm operates.

The dominant market share of the multiple retailers in most countries means that a large producer cannot afford to disregard their business, and must try to gain a product listing with one or more. Since they themselves continually seek products which will differentiate their offer and assist their competitive struggle with other multiples, they are generally willing to listen to presentations from specialist suppliers. If convinced, they will then invariably provide considerable practical support for a new supplier. It is therefore a sound strategy to identify a multiple retailer which actively seeks such producer alliances or marketing relationships. For smaller producers this is probably only achievable through group marketing, which will guarantee sufficient product volume and year-round availability, or by own-label manufacture.

Product range

Most sectors have distributors who concentrate on a very narrow range of products, and others who handle a range consisting of tens of thousands of items. Specialist wholesalers and retailers are particularly common in some food product groups (bakers, butchers, organic and health food stores), and they still dominate the marketing system in some countries which are feasible export markets (Greece and Italy, for example). In other countries their presence is negligible: for instance, the USA.

The chief advantages to the small business of a specialist distributor are:

- specialist product knowledge
- lower requirement for promotion/advertising, since customers encounter the product in their normal use of the retailer, and the fact that it is stocked is an endorsement
- quantity requirements may be smaller (certainly than national multiples).

The potential disadvantages include logistics difficulties if widespread distribution is desired, but this may sometimes be overcome by using specialist *wholesalers*. More seriously, in countries dominated by one-stop multiple shopping, sales may be restricted to consumers who are prepared to go out of their way to a specialist shop for certain purchases.

Multi-product firms with a wide product range increasingly have a product depth in many sectors which may match that of specialist distributors. This is indeed becoming the norm as multiple retailers and some wholesalers seek to establish a particular image in the sale of some product groups. Some seek a reputation in wine, for example, some in fresh or organic produce, cheese, delicatessen products, bakery etc., and potential suppliers may concentrate their selling effort accordingly. The same is also true in the non-food sector, where wide-range outlets actively seek to differentiate their customer offer through a specialist reputation for (say) handcrafted, 'traditional' or other distinctive products.

Selling method

The selling method practised in a channel may be relevant to sales promotion and the introduction of new products – particularly if these need to be explained and positively sold. The major distinction is between customer self-service ('cash and carry' in the wholesale situation) and service by sales staff, since the latter provides an opportunity to draw customers' attention to new lines, make recommendations, and encourage sampling. In the self-service situation the product has to speak for itself, so the customer's attention has to be drawn to it before entering the premises (by media advertising and other promotion), and/or the product itself must attract attention at point of purchase (packaging, merchandising, price promotions).

In most countries the distinction between self-service and serviced selling is disappearing. At retail level, self-service by the consumer is normal for the vast bulk of food products sold, supplemented by service counters for specialist product sectors in which some customers value individual attention and advice (usually meat, fish, cheese, delicatessen). Even in the self-service store the opportunity therefore exists to supply customer information via trained sales staff. At the wholesale level this opportunity is rarer, but in cash-and-carry outlets where some trained staff are employed (typically in butchery) a training and promotional input may be possible if it complements the wholesaler's own trading style and objectives.

Terms of trade

The terms of trade on which sales are to be made will be the object of negotiation between supplier and potential buyer, and will include:

- detailed delivery arrangements (place, time and form)
- sale or return
- warranties/replacement policy
- agreed services (most commonly, in-store merchandising by the supplier, joint advertising or promotion), the cost of which must be built into the evaluation.

A major consideration will also be the payment period. In some countries (including most EU countries) this is regulated by law and relatively short: typically 15-30 days. Where this is not so, the disadvantage of a long payment period may outweigh other positive factors which favour a particular distribution outlet. For example, whereas small retailers generally pay within 7 to 14 days post delivery, some large firms delay payment for as long as 60 to 100 days, which places a major strain on the supplier's cash flow.

Reputation

The reputation of a potential distributor is clearly relevant to sales potential and market prospects. For public companies, useful indicators will be found in published information: annual reports, trade and media articles which report market share, profitability and other financial performance measures. Equivalent information about other distributors will rarely be available, and in both cases there is no substitute for personal research, focusing on observable factors like quality of customer service, retail environment/style, promotional style, innovativeness etc. A business reputation for fair dealing will also be important, but again, there is no substitute for personal assessment – which is where the actual sales negotiation process provides an under-rated opportunity.

GAINING ACCESS

Gaining retail access for a product – even gaining the opportunity to introduce the product to the potential buyer – is much more difficult than identifying the distributor. This is true whether the distributor is a small independent supplier or a large retailer multiple. The latter may literally be more difficult to access in the sense that the system is difficult to penetrate, but the independent retailer who is often an owner-manager has pressures on his time which make him as reluctant as most farmers are to waste it on 'rep' visits. Both large and small distributors will therefore be intolerant of inappropriate sales approaches, so preliminary thought and planning are essential in both cases to target the right buyer and present an appropriate sales 'pitch'. The efficiency and style of the first contact will also invariably be critical in gaining attention in an overcrowded marketplace.

Persuading a distributor to buy and stock a product generally requires personal selling by the producer or by sales personnel, usually by face-to-face interview with a buyer or the buying department of a large company. This is an unrivalled opportunity to explain the distinctive attributes and particular benefits of a product to the customer, and to establish confidence in the supplier's ability to deliver. Producers who have done their own market research and product development should therefore be their own best salesman, since they have all the relevant information at their fingertips.

In practice they commonly fail to maximise the opportunity because they are intimidated by the encounter, and either over-diffident or over-enthusiastic about their product. Many under-estimate the professionalism required, forgetting that professional buyers are also professional sellers, who will have no time for a badly judged sales approach. This is particularly true of the highly competitive, congested food product market, where an unprofessional sales approach may cast doubt on the product, and the supplier is unlikely to have a second chance. The sales situation also provides an opportunity – possibly the only opportunity a supplier has – to weigh up the *buyer* to discover what he is like to deal with, and whether he is worth targeting.

PRODUCT PRESENTATION

Product presentation demands the same careful planning and preparatory work as any other part of the marketing process, plus presentational skills which can be learned: salesmen are made as well as born. To save their own time some large companies actually provide guidance on an appropriate approach; more commonly the onus is on the supplier to design an appropriate sales approach.

The data required for the product presentation is that already used for the product specification (distinct product attributes, price, consumer and customer data, knowledge of the market and the buyer's competitive strategy, logistics, the supplier's own strengths and ability to meet marketing schedules etc.). The main danger for the entrepreneur-producer is to avoid information overload in communicating the product offer. It is therefore worth recognising the stages which professional salesmen observe in their product presentation (Table 10.2): first gain *attention* and *interest*, then stimulate *desire* and *action*.

The critical stage in a congested market is the first contact, when the buyer's attention and interest need to be gained amid the surrounding noise of other product offers. This may be an initial telephone contact or written approach which must be pertinent and to the point, identifying immediately and concisely the innovative quality of the offer and/or its 'distinctive asset': guaranteed traceability; hand-crafted; organic etc.

Once attention is gained, the buyer's interest must be stimulated by providing as much technical information as he requires about the product and its uses, input materials and processes. Most importantly, the buyer's perceived risk in trying an unknown product must be minimised, by stressing its particular appropriateness for the buyer and its competitive advantage over other products already known (which is where the value of good distributor and competitor analysis becomes evident). Guarantees will also invariably be required at this stage to reduce the buyer's risk (sale or return, unconditional product replacement), together with some promotional back-up.

The way in which a sales interview is handled is also instrumental in gaining sales, since confidence in the supplier as well as the product must be established. An understanding of buyer behaviour is invaluable in gauging an appropriate style of presentation as the interview proceeds, to maximise the opportunity to impress and minimise the opportunity to annoy or offend. The style can to some extent be predicted on the basis of information about the buyer's own retailing style and attitudes. In the interview situation, however, a good salesperson will rapidly assess the personality of the actual buyer and respond to signals which indicate objections, doubts and signs of data overload, to which the presentation is instantly adapted.

ATTENTION	*Prospecting* *Pre-approach*	Identify potential customers Select targets Research/plan sales pitch
INTEREST	*Approach* *Presentation*	Establish style, credibility Highlight product attributes, competitive advantage Demonstrate, sample
DESIRE	*Overcoming objections* *Negotiating* *Closing deal*	Anticipate objections, rebut Offer incentives, guarantees Discuss terms
ACTION	*Follow-up* *Feedback*	Deal efficiently with delivery and documentation, communicate Suggest other products/services Accept and act on criticism, suggestions

Table 10.2 Stages in the selling process (AIDA)

For the novice salesman, textbooks provide 'buyer typologies' which suggest visual and behavioural clues from which an appropriate approach may be deduced (Table 10.3). These may seem naive, and like all such generalisations they are very fallible. (To state only the most obvious objection arising from Table 10.3, interviewers may be in someone else's office, an interview room, a store-room or a shop which offers no personal clues whatever). The types are nonetheless sufficiently recognisable to suggest the kind of detail which can be turned to advantage in the face-to-face interview. For example, from the information in Table 10.3 the following conclusions may be drawn which are helpful in approaching different kinds of buyer (company).

- *Analytical buyers* are cautious, and seek to reduce risk by attention to detail and careful analysis: they therefore want to know everything about a product and the producer. This may be demanding on the supplier, but since their decisions are based on logical analysis they are open to rational persuasion and hard data, and once convinced they will be loyal buyers for a reliable supplier. Most multiple retailers and their individual buyers are 'analyticals', because they are conscious of their public image and the need to justify their activity, and this behaviour is reinforced by the competitive nature of the marketplace and public health risks attached to food retailing.

- *Drivers* are forceful, dominant personalities who like to be in control, and are more concerned with results than means; they are thus often 'bottom-line' and 'time is money' people. Faced with a range of alternatives they will make quick decisions, but they are typically uncompromising in what they want. The operators of discount stores and small retailers are often classed in this group.

- *Expressives* are 'big picture people': dismissive of detail, seeking the best solution, not a rehearsal of alternatives. They respond with enthusiasm to new ideas and may make quick decisions; they are consequently an attractive option for a new supplier, but are impatient of detail and explanations if things go wrong.

ANALYTICALS	DRIVERS
Computer print-outs/charts visible	Seem to be in permanent crisis
Little office decoration, 'all-business' atmosphere	Office littered with paper, disorderly
Order and neatness in office, personal appearance	Big desk, executive armchair
Technical background detectable in vocabulary	Constantly on and off the phone, up and down
Businesslike manner, polite, not effusive	Polite but challenging
Conservatively dressed	Conservatively dressed, jacket off

EXPRESSIVES	AMIABLES
Showy office, manner	Orderly office, unshowy
Office and desk cluttered	Photos of children, wife, personal mementoes
Desk placed for contact	Warm, friendly, comfortable atmosphere
Expansive gestures, arm waving	Slow spoken, quiet manner
Fashionable/individualistic dress	Desk placed for contact, easy armchairs
Exaggerated speech	Quiet dress, welcoming manner

Table 10.3 Buyer types as indicated by visual/behavioural clues

- *Amiables* like to build relationships, so may take their time getting to know a potential supplier, but the resulting relationship will last indefinitely and be difficult for competitors to break. This provides security once access is gained, but both partners may suffer if the 'cosy' relationship ignores changes in the marketplace.

In business as in life these categories obviously merge, but in prevailing market conditions the typical buyer is held to be an 'analytical-amiable', reflecting a move away from confrontational selling towards relationship marketing.

FOLLOW-UP SELLING

Once a successful initial purchase has been achieved, two situations can usefully be distinguished.

The *straight rebuy* ideally becomes habitual ('low involvement'), and requires little or no further intervention by the seller or buyer beyond routine re-ordering, delivery and payment. (For example, the routine relationship involved in doorstep milk delivery, or dairy to wholesaler or retail store). Many retailers have computerised systems which automatically trigger re-ordering when stocks fall below a predetermined level (*vendor-managed inventory*); smaller businesses expect a salesperson to call or phone for an order, or to re-stock the in-store display (*merchandise*) and collect payment.

The *modified re-buy* situation arises when a modified product is introduced: new flavours, varieties, added value packs, new model etc. On this occasion the risk to the buyer is reduced, so less information and support are required, but product innovations must always be drawn to the buyer's attention. Where mutual confidence exists this information process may itself become routine (for example, a simple advice of the planned change, new prices, delivery details etc.).

The establishment of a routine trading relationship should be a high priority since it reduces the time and effort involved for both parties. The basis for this must be a thorough understanding of each party's objectives and operational needs, and at the beginning the onus is invariably on the seller to know and adapt to the buyer. (The customer may not always be right, but he expects to be treated as such.) As the relationship and mutual trust grow, a satisfied customer will generally be prepared to modify trading arrangements to suit the seller's needs, provided they are well-explained, justified, and not damaging to the buyer's interests.

DISTRIBUTION MANAGEMENT

Distribution management involves the efficient supply of product support and customer services. Product support entails sales-related services: promotion, advertising, product improvements. Customer services include all the logistics necessary to make the product available to point of retail sale (Figure 10.2): stock holding (*inventory control*), efficient delivery, data management (order processing, order status information, invoicing), which is increasingly achieved by electronic data interchange (EDI). These services entail further activities: protective packaging, standardisation of pack, bulk loads etc. All these are expensive undertakings – typically 20 per cent of the retail price, and their management requires constant attention to detail because any failure could result in distributor de-listing and lost sales. Even more seriously, persistent failure could establish a reputation for unreliability resulting in long-run damage to the business.

The key to logistics management is to deliver the required level of service at minimum cost. The service level need not be the maximum obtainable since this may be too expensive and unnecessarily high, but customer service should not be sacrificed simply to reduce costs. The priority is therefore to identify exactly:

- what the customer requires (for instance, frequency of delivery: daily, weekly, monthly)
- the service level provided by competitors
- the cost of supplying the customer and beating the competition.

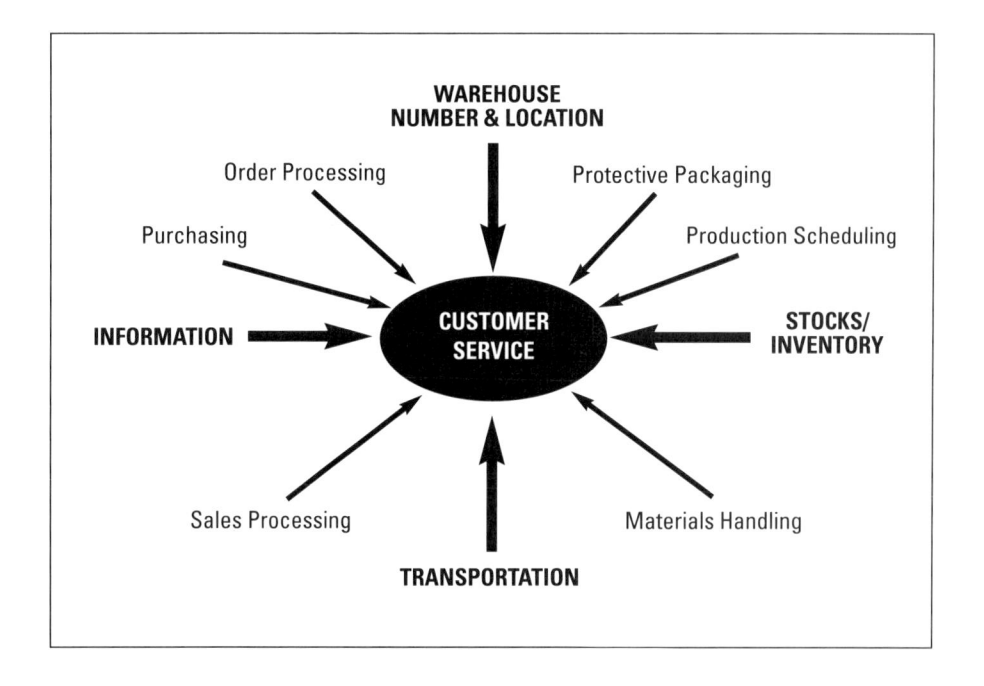

Figure 10.2 The logistics mix

The means by which this is achieved may not interest the customer, in which case the decision will be made on a cost basis, and one activity may substitute for another. For example, highly efficient manufacturing plus rapid delivery methods may allow stock levels to be minimised. High stock levels may allow a product to be manufactured in batches once a week, but delivered every day. It is also possible to substitute one means of carrying out an activity for another. For example, fast but expensive air-freight may be substituted for slow but cheaper road/rail. Where road transport is required, hired/leased vehicles, own fleet, or haulage contractor all need to be considered because all may be capable of achieving the customer's availability requirements. However, each has its own cost structure and risk of failure, which must be evaluated.

Effective distribution management requires constant monitoring of the business's own performance, as well as that of the distributor. Internal appraisal should be achieved by systematically benchmarking internal performance by reference to promises, customer expectations and satisfaction, and competitor performance – the last two researched by the salesperson or merchandiser through observation and questioning of distributors and consumers. This will identify where remedial action is needed to improve performance, and if performance is better than the competition, where cost savings may be possible.

Distributor performance will be assessed primarily by sales achieved. If sales are unsatisfactory and there is no logistic failure, the reason must be identified. It may be that distributors 'push' competitor products because they make a bigger contribution to their own costs. Ways of increasing the percentage contribution should therefore be sought, or other forms of sales incentive explored (for example, increased advertising, better merchandising). Other reasons for poor distributor performance may include defective location, product range, retail management, low stock levels – all leading to unsupplied or dissatisfied consumers and lost sales. The precise reasons must be identified, however, and remedial action taken.

Where distributor inefficiency is at fault, appropriate action may include training or persuasion, or alteration of the terms of sale. If this fails, or the aims of the two parties are at odds, a change of distributor will be required. For example, a producer may see a product as a high-quality premium offer commanding a high price in a niche market, but for any number of reasons a distributor may disagree (for instance, change of management or ownership). Where this arises, other distribution outlets must be sought – which may include the option of direct sale.

DIRECT SELLING

Direct selling is usually understood to mean the use of retail premises and house-to-house sales, but it also includes a rapidly widening range of 'direct marketing' techniques which include:

- mail order
- computer and TV home shopping
- telemarketing (telephone 'cold calling' sales)
- freephone invitations
- telephone and electronic business/service directories (for example, Yellow Pages and the Internet)
- direct response coupons delivered to the door, with increasing precision of customer targeting
- franchising – a partnership between a producer and a salesman which exploits their respective skills and resources.

In all situations except the franchise, the producer is the principal in the sale, and ownership of the product passes direct to the consumer without any intermediary party (except own sales staff). In some situations personal selling is involved, and its importance to enterprise viability is such that any reader venturing down this path should seek detailed advice on selling techniques and sales management from a specialist textbook.

DIRECT MARKETING

Direct marketing (mail order, sales via TV and Internet etc.) does not currently involve personal selling because the product presentation is non-interactive, and customer questions cannot generally be answered. The opportunity to 'customise' the offer is therefore more limited. Product information is generally pre-programmed, and products are simply ordered or not on the basis of the information supplied. This is a rapidly evolving area, however, and developments in information technology and consumer acceptability of the medium will certainly introduce interactive communication which could present new opportunities for small suppliers. The problem of remote location and physical access, for example, could be largely overcome; local suppliers may also find a niche as suppliers for electronic ordering services or even retailer multiples seeking to establish a distribution network for specialist products.

In the absence of the personal contact, direct marketing initiatives rely heavily on effective communication of product information through written or electronic media; careful choice and presentation of product information are therefore essential. In direct marketing the viewed product and its setting are a substitute for sales people, so great care is necessary to ensure attractive presentation; presentation and reproduction must also be accurate, however, because there is a legal obligation in most countries to provide accurate trading descriptions and products which match the description. Since the customer is still unable to assess size, quality and other attributes before purchase, money-back guarantees and unconditional replacement must also be supplied. These are expensive to operate, but sales will be disappointing without such guarantees, and the media may refuse to carry advertisements which do not offer them.

Direct marketing has the advantage of eliminating the need for retail premises and face-to-face contact, and makes logistics extremely simple – though it is of course dependent on delivery systems outside the supplier's control, which may be open to disruption: for example, mail strikes, road and transport delays. Costs associated with premises and distribution are thereby reduced, but substituted for by much higher advertising and promotion costs.

For many small businesses, the small-ad columns in newspapers or local radio and TV advertising slots may be a very effective and cheap way of selling the volumes of product available. If larger volume sales are required over a wider area, much greater expense will be incurred: for instance, mail order speciality foods sold via glossy magazines and weekend newspaper supplements. The publishers of these magazines have good readership information so it is possible to target niche markets fairly accurately, and although these advertisements are expensive, they may be highly cost-effective. This is even more true of the specialist press which can be used to target a very small niche: for example, readers of recreational, hobby and special interest magazines.

FRANCHISE SELLING

A hybrid form of direct sale is the franchise, where a producer (franchiser) and an independent retailer (franchisee) enter into a sales agreement which pools some of their respective responsibilities. The franchiser effectively recruits the entrepreneurial sales skills of the franchisee, usually within a standard presentational formula which is capable of replication to additional outlets as demand allows. On a large scale, a chain of retailers (with or without retail premises) is created which is dedicated to the sale of one manufacturer's products; each outlet is owned and run by people with sales skills whose return is directly related to sales achieved. Products may be supplied on a sale-or-return or a no-sale-no-pay basis, and the producer may assist with the capital cost of premises; he will certainly assist in promoting the brand.

The best known examples of franchises are international operations with household names like Benetton and McDonalds, and the method is particularly common in the fast food catering sector (themed restaurants and take-aways). However, franchising on a small scale has been successfully adopted by a number of small farmhouse producers – typically, the opening of ice-cream and milk parlours retailing farm production. There is no reason in principle why it should not be more widely adopted, but specialist advice will be required in drawing up the legal agreement between the parties.

CONCLUSION

Selling to trade customers and consumers is a crucially important marketing function, particularly where personal contact is established between buyer and seller. In industrial markets this may be the only contact between buyer and seller, and the only opportunity for active product support and sales promotion. At both consumer and industrial levels, the sales situation provides a regular point of contact for product introduction, demonstration, information provision, and potential added value through added services. It is also an unrivalled opportunity for continuous customer feedback which is a major source of product and business development, and a strong component of the promotion mix to which we now turn our attention.

CHAPTER 11

Promotion

The myth that good products sell themselves is precisely that: a myth. However good a product is, potential customers need to know that it is on the market, and customer interest has to be transformed into product sales. This requires effective communication of information about product attributes, price and availability, and promotional activities which persuade customers to buy.

Everything which a business does communicates something to the customer, whether or not it is intended: product packaging, premises, vehicle livery, business manner and style, personal delivery of service products etc. Together these create a corporate image which can and does influence demand; promotion sets out to ensure that the image is positive, not negative, and encourages purchase.

This should not be confused with the popular notion of the 'hard sell', which entices customers to buy unwanted products. Promotion simply seeks to ensure that customers choose one product rather than another by stressing the particular benefits it offers them. This is necessary to gain and maintain customer awareness of a product in the presence of many competing offers, and to overcome habitual buying behaviour: it is thus an essential part of a segmentation strategy, and a major management tool in the drive for competitive advantage and sales.

The promotion mix is the combination of communication methods which a business uses to achieve this: hence the alternative term 'communication mix'. Four broad categories of promotional activity are normally recognised (Figure 11.1): paid advertising – the most visible form; unpaid publicity; personal sell-

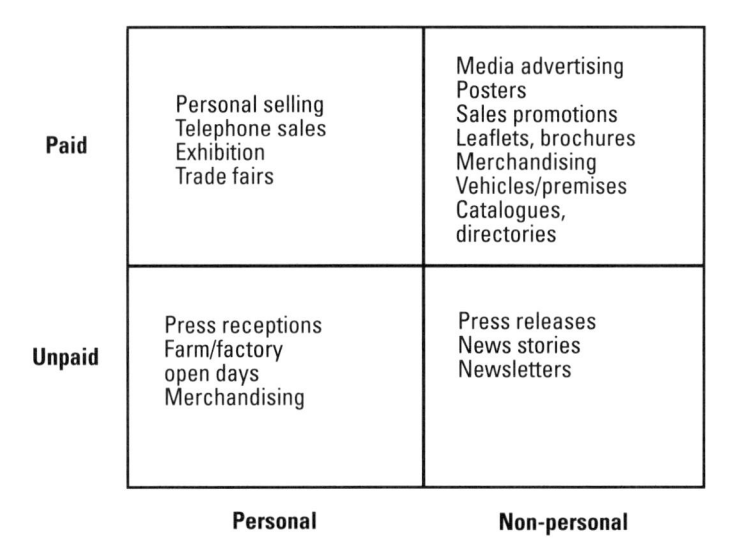

	Personal	Non-personal
Paid	Personal selling Telephone sales Exhibition Trade fairs	Media advertising Posters Sales promotions Leaflets, brochures Merchandising Vehicles/premises Catalogues, directories
Unpaid	Press receptions Farm/factory open days Merchandising	Press releases News stories Newsletters

Figure 11.1 Promotion activities

ing, and sales promotions. The role of personal selling was considered in the last chapter, and sales promotions were mentioned in Chapter 8 as a pricing tool: this chapter introduces the remaining promotional tools, and the management considerations involved in the design of an effective promotion mix.

The right balance of promotional methods depends on the individual business, but the same principles apply whether it is an owner-managed business spending a small sum on targeting a relatively small number of customers, or a large company or a producer group involved in joint promotion, where the budget may have alarming proportions. The design of the mix draws on the customer data generated by market research (Chapter 6), since the promotion message encapsulates the product offer. The problem is to get the right message to the right target audience via the right medium at an affordable cost.

This is not easy in the presence of noise from surrounding competitor offers and the customer's own preoccupations. It is also very easy to spend money on promotion without any very clear conception of the realisable gains, or even whether the expenditure and effort were effective. Promotion is consequently one area in which a professional input is not only advisable, but invariably necessary to obtain technical expertise which ensures that effort and expenditure are maximised.

Many small businesses admittedly undertake their own promotion successfully without professional help, but they generally buy in some technical skills – usually design and execution of artwork for promotional material, interior design for business premises etc. Many of them therefore conclude sooner or later that it is sensible to employ an agency to oversee the execution as well as the design of the promotion mix. This gains the benefit of insider knowledge and associated economies: for example, access to market research data, preferential rates for paid media, access to prime print and poster sites or times in broadcast media. It therefore achieves both greater effectiveness and *cost* effectiveness than self-provision, and frees management time for production and other marketing activities. Individual promoters and members of marketing groups still need a knowledge of promotional objectives and techniques, however, to evaluate the appropriateness and effectiveness of campaigns commissioned and funded on their behalf.

THE PROMOTION MIX

A business will build its promotion mix on a combination of the activities shown in Figure 11.1, one of which may be, but is not necessarily, advertising. The mix will vary with the message and the medium, the product, the customer, the competition, the available budget, and the business and promotional objectives, and it will vary over the product life cycle.

In the case of service products, promotion substitutes for the distribution of physical products; advertising and personal selling therefore figure prominently in the mix. These are both 'above the line' items which entail payments to media suppliers (newspapers, radio, TV), fees to an advertising/promotion agency, and labour costs (sales staff, own time). Consequently, for service products the promotion cost figures prominently in the feasibility appraisal, and efforts to reduce it may be counterproductive. A lower advertising budget may simply bring fewer sales; cheaper promotional literature may convey a poor impression of a service product. The high cost of high-quality advertising copy and promotional brochures may therefore be a necessary trading cost to communicate the intangible product in a tangible form.

In all situations the primary factor determining the communication mix is likely to be objectives combined with cost. Communication involves a message and the means to deliver the message, which can only be determined by reference to objectives sought. For example, trade fairs are an excellent way of meeting trade customers or seeking international distributors, and they are comparatively low-cost relative to the business exposure achieved. Sales promotions are good at generating short-term sales growth but they may be costly to run; they are also inappropriate to the long-run objective of establishing an image as an 'ethical' supplier (how a business is preserving the environment, is kind to animals, or is socially responsible). In this case public relations (PR) linked to media publicity is more effective and less costly.

OBJECTIVES OF PROMOTION

The short-run objective of promotion is to achieve sales by gaining a high degree of customer recognition for the product in competition with comparable offers. The long-run objective is to achieve a customer base loyal to the product and the supplier, which reduces the marketing effort and cost, and provides a basis for sustainable trading. These twin objectives are achieved by a combination of:

- product-related activities which enhance the core product and identify it with the supplier (packaging, premises, vehicle livery, plus 'intangibles' like presentation, service, business style etc.)
- short-run sales promotions and associated advertising and PR
- long-run advertising and public relations activities which build a corporate image and habitual buying behaviour through brand loyalty.

The product message and the corporate image to be conveyed must be clearly determined, credible, and deliverable. It must also be consistently communicated over time and through every business activity if expenditure on well-conceived and executed promotion is not to be put at risk by management inattention.

Branding

All this implies product branding, since it is impossible to promote an undifferentiated product or an anonymous supplier. For producers of undifferentiated agricultural commodities this was only achievable in the past by generic advertising for an entire product sector, which has been very effective in sustaining demand in sectors where adverse publicity and competition from substitute products threatened to reduce it. Generic advertising is in other words a necessary production cost, even though producers of subsidised commodities fail to see it as such (Box 11.1) . Producers of unsubsidised agricultural and horticultural products, by contrast, have been much more active in promoting their output since their revenue depends on it.

One valid criticism of generic promotion is that it benefits all producers of a promoted product, irrespective of the quality produced or the merit of individual suppliers (or regardless of financial contribution where levies are voluntary). This has been avoided by the establishment of producer groups formed to establish their own brand, which is backed by production and marketing disciplines to guarantee a standard product of known origin and quality. The spill-over to other producers is thereby minimised, and the individual's investment in production effort and financial support for the brand is repaid in the form of secure market access and, often, a price premium.

Producer reluctance to pay for beef levy may be false economy

No-one disputes the need to promote British beef. The question is who should pay? Farmers Weekly readers have a clear answer, judging by our recent telephone poll. Nearly 80% of callers were opposed to any increase in cattle levies to fund promotion ...That could be an expensive mistake. Without contributions from producers the future of UK beef promotion is bleak ...
The government will allocate £2.5 million ... to help the MLC and its Northern Ireland counterpart ... But even optimists consider that extra cash from government is extremely unlikely. Yet MLC Chairman Don Curry has said his organisation needs an extra £15 million in the next two years to mount a high-profile campaign to recover the 20% fall in beef consumption caused by BSE.

Farmers Weekly believes producers have a responsibility to contribute to campaigns which benefit their businesses and a meat industry worth £17 billion. True, beef producers have been hard hit by the catastrophe that is BSE, but there is no doubt that such promotions provide excellent value for money. Last year the MLC's Quality Mince Mark initiative, which saw the blue quality mark prominently displayed on mince packaging, returned £35 million in extra sales from an investment of £3.5 million.

Meanwhile, some consumers' attitudes to beef are hardening. We need to win them back sooner rather than later to safe, high-quality beef produced from animals raised under exacting welfare standards.

The future of our industry is at stake. There should be no beef about a moderate increase in cattle levies.

Farmers Weekly Editorial, 14 February 1997

Box 11.1

This does not invalidate the need for generic promotion; on the contrary, the synergistic effect achieved by joint promotions maximises the total spend and value for money. This is essential in the face of massive expenditure by private firms and other product sectors which are all competing for a larger share of a relatively static demand. The same objective is also now achievable through producer alliances with retailers who are increasingly conscious of the benefit of a named supplier alongside their own label, as an additional demonstration of traceability from farm to plate. This has significantly expanded the opportunity for producer labels to achieve a retail market presence.

Once a product has an effective brand identity, it is possible to deploy the entire armoury of promotional tools to maintain its market presence and gain customer loyalty. Two approaches should be distinguished, since they have different media and cost implications and different tactical requirements:

- a 'pull' strategy targets promotion direct to consumers, 'pulling' demand into retail stores; the main promotional tools are therefore advertising and direct marketing
- a 'push' strategy persuades distributors to stock the product, and relies on personal selling and trade promotions to push supply into wholesale and retail outlets.

In practice the two are normally used together: consumer advertising to pull demand is used alongside trade promotions to push supply simultaneously into retail outlets (Plate 18). The combination of pull and push will generally be resource- and channel-related. For instance, advertising is expensive and cannot be undertaken without a brand, so a product must be identifiable: hence the need for generic promotion of undifferentiated commodities such as milk, beef, lamb, pork, etc. In a marketing alliance, one partner commonly provides one component of the mix (media advertising, for instance), while another provides sales support via price promotions and in-store merchandising. Effective coordination is therefore essential to ensure that the same message is being communicated throughout. The potentially high cost of promotion also means that careful planning is necessary to ensure that objectives are achieved as cost-effectively as possible.

Growers' identity is revealed

There are continuing signs that multiples are realising the public relations value of having individual growers' identity on produce alongside their own brand. Until recently, own-label suppliers ... have remained anonymous. Opening a new onion store ... Tesco's produce trading director said 'Our customers like it. Not many growers would put a poor product in a pack and put their name on it.' ... Another case of improved identity on a regional basis is also becoming evident, with multiple acceptance of the Cornish King vegetable brand developed by ADAS. First used on potatoes, it has now been extended to cauliflowers and spring greens.

The Grocer, 4 January 1997

Box 11.2

THE COMMUNICATION PROCESS

Effective communication depends on transmitting a clear and informative message which is rarely as simple as 'buy product X' – though admittedly that is the underlying objective. The message must communicate what is essentially different about a product and the particular benefits the customer will derive through purchase: its *unique selling proposition* (USP). This generally has to be achieved in a very short time or space because paid media can be very expensive, and the customer's attention span is limited both by the ephemeral nature of most media and the low-priority they give to advertising.

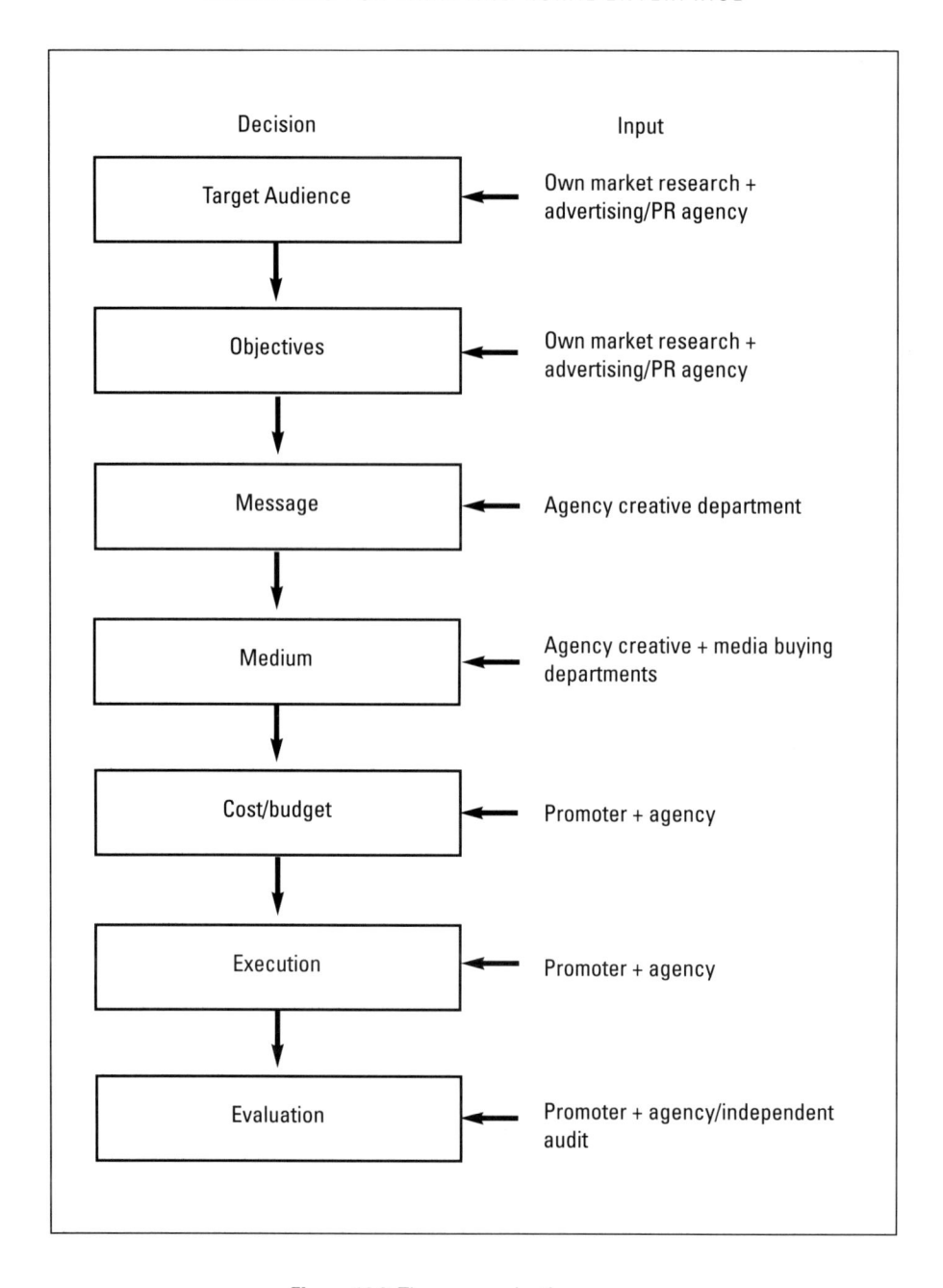

```
        Decision                    Input

    ┌─────────────────┐      ← Own market research +
    │ Target Audience │        advertising/PR agency
    └────────┬────────┘
             ↓
    ┌─────────────────┐      ← Own market research +
    │   Objectives    │        advertising/PR agency
    └────────┬────────┘
             ↓
    ┌─────────────────┐      ← Agency creative department
    │     Message     │
    └────────┬────────┘
             ↓
    ┌─────────────────┐      ← Agency creative + media buying
    │     Medium      │        departments
    └────────┬────────┘
             ↓
    ┌─────────────────┐      ← Promoter + agency
    │   Cost/budget   │
    └────────┬────────┘
             ↓
    ┌─────────────────┐      ← Promoter + agency
    │    Execution    │
    └────────┬────────┘
             ↓
    ┌─────────────────┐      ← Promoter + agency/independent
    │   Evaluation    │        audit
    └─────────────────┘
```

Figure 11.2 The communication process

Figure 11.2 indicates the process involved in communicating an effective message, and the stages where a professional input is likely to be required and most effective. The promoter generally determines the objectives (sales generation, quality reputation, new product launch etc.), and identifies the target audience on the basis of market segmentation data. Specialist advice from a PR or advertising agency is generally necessary to devise the best message, to identify the best medium to communicate with the target segment, and to cost the programme. The agency may also execute and manage the programme. The evaluation and measurement of programme effectiveness may then be undertaken independently of both, by a third firm which is a specialist in the job.

TARGET AUDIENCE

The likeliest target audience for promotion is existing and potential customers (consumers and intermediary buyers), but a business may also wish to communicate with the general public as part of a PR effort to build its corporate image. This is more commonly done by producer groups and trade associations, but the same elements of the mix, the same techniques, and the same principles apply in both cases.

The primary focus of promotion will be the target consumer group identified by the market research. During the product lifetime secondary groups generally emerge who are found (from new research) to fall outside the boundaries of the initial segment, and are purchasing the product for the purpose/satisfaction originally foreseen or have found other uses/satisfaction. These may constitute sufficiently large secondary-use markets to make a separate communications effort worthwhile. Trade customers require separate targeting, as do a wide range of other organisations which can pass on information about a product: importers and exporters, tourist boards, special interest groups etc.

The nature and extent of each target audience's existing knowledge and attitudes towards a product and a supplier must be identified so that promotion is well targeted. For instance, promotion which is intended to create awareness or interest in potential non-users will need a different information input, and use different means to gain purchase, than promotion designed to change adverse perceptions or attitudes of existing customers.

COMMUNICATION OBJECTIVES

The commonest objectives of promotion are:

- to increase sales
- to stimulate awareness and trial of a new product in advance or at time of launch
- to stimulate awareness and credibility of a business as a basis for other trading activities
- to reinforce a positive image of a product/business/brand
- to remind customers of the existence of a product/supplier/brand
- to build/reinforce customer loyalty by stressing track record, trustworthiness of supplier, safety and traceability of products
- to change attitudes to a product/business/brand.

Most businesses seek to achieve several or all of these in a single campaign in which different media and promotional tools are used to achieve different objectives simultaneously: for example, media advertising to heighten business/brand awareness; sponsorship of a charity/local school sports event; sales promotion to encourage immediate purchase; money-off coupons to stimulate repeat purchase, etc.

Sometimes it is not easy to reconcile short-run with long-run objectives. For instance, it may be difficult to urge distributors to stock higher product levels and consumers to buy more when a long-run campaign is simultaneously stressing the 'partnership' qualities and social responsibility of the supplier. Some objectives are also much more difficult than others to achieve. It is more difficult to change negative customer attitudes and habitual behaviour than to establish a positive image where there is none at all (for a new product or in a new market, for example). It also requires a long-run horizon, because attitudes and habits are difficult to change, and once an attitude is changed promotion will have to be sustained – perhaps for years – in order to maintain the improved perception. This may be difficult to fund, but may be an unavoidable cost of trading.

MESSAGE, MEDIUM, COST

Whereas target audience and objectives are decisions for the promoter, message and medium should ideally be partnership decisions taken in association with a professional. The two decisions – what to say and how to say it – interact closely. The adage that the message is more important than the medium is rarely true in marketing, where the right message in the wrong medium is a waste of money. Some media also result in distortion either in the transmission or receipt of a message, so for some messages (public health and safety aspects, for example) a non-ephemeral medium is required (printed material, not TV advertising).

The choice of message and medium is strongly influenced by the available budget. Media differ widely in their cost, and this must be considered when the message is being designed. Costs are associated with origination (creation and design, artwork, interior décor etc.) and with subsequent exposure/transmission (space costs in newspapers, air time on TV or radio, refurbishment of premises). These must be clearly distinguished because they have different budget implications. For example, the origination costs for a price discount coupon scheme may be minor (design + publicity or advertising), but the redemption cost of coupons will be high, and an ongoing cost for the duration of the redemption period. The origination cost of media advertising will also be relatively low compared with recurrent transmission costs. Other promotional activity will incur high initial cost (shop fitting or liveried vehicles) but subsequent costs will be comparatively low.

The calculation of total costs is needed in order to establish whether a promotion is affordable. Whether the investment is worthwhile is related to effectiveness, which is difficult to measure. Cost may be expressed in terms of cost per opportunity to view or register the message (exposure), which allows the comparison of an inexpensive but localised programme with an expensive but large-scale one: for example, the £200,000 cost of a magazine advertisement compared with the £1 million cost of a television advertisement. The cost of the television advertisement may have a lower cost per exposure than a £200,000 magazine advertisement, but this is irrelevant if the business cannot afford it. Equally, television may be an inappropriate medium for the message which needs to be communicated.

Message	TV/Radio	Press Trade Fair	Sales	Publicity	PR	Personal	Leaflet	Exhibition
New convenience food	✓	Consumer	✓					
Value for money with multibuy	✓	Consumer	✓					
New product	✓	Consumer	✓	✓				
Launched in 4 months	-	Trade	-	✓	✓	Trade		✓
Will solve your problem	-	Specialist	-	✓		✓	✓	✓
Complicated but good: send for details		✓		✓			✓	
For sale		Small ads					✓	
Now open/available	✓	✓		✓	✓			
Good for you, so change your lifestyle and buy		✓		✓	✓		✓	✓

Figure 11.3 Matching medium to message

Choosing the right combination of message and medium will therefore normally be an iterative process, since the ideal medium may too costly, and the message has to be modified for another medium so that it can be afforded (Figure 11.3). There are also many free promotion opportunities which may be equally effective, and will reduce the need for or reinforce the impact of paid promotion.

MESSAGE

The message is normally conceived by the promoter in relation to the required product attributes identified by market research (ease in preparation, taste, natural, quiet weekend in the country etc.) and the appropriate style to suit the target audience (factual, emotional, challenging etc.). The advice of an agency will be needed to refine the message and its expression, in order to maximise the effectiveness of the communication within the very limited exposure space/time available. For example, the message should be action-oriented to prompt customer response (pick up a brochure, visit this leisure facility). The source and identity of the promoter and the product must be clear – hence the value of a visual brand/logo which instantly triggers recognition, and may make explicit identification of the source unnecessary. Any ambiguity must also be avoided which might result in misinterpretation and possible disappointment or offence.

The ephemeral nature of most promotional media puts a high priority on the ability to communicate information in a form which attracts attention and retains interest long enough for the salient details to be absorbed. The AIDA model introduced in Chapter 10 is helpful in suggesting different kinds of information and style for different stages of the attention-interest-desire-action process (Table 11.1).

ATTENTION	Two penguins talking	Unusual/amusing association of ideas, eye-catching
INTEREST	A chocolate biscuit with a cash-off offer	Retain attention by adding information relevant to customer
DESIRE	Low fat, easy-open pack, resealable for freshness	Appeal to psychological benefit, special product attributes
ACTION	Available now in your local store, limited period, so hurry out and buy	Facilitate purchase, motivate purchase

Table 11.1 The AIDA model applied to communication message

Content

The content of a promotional message may be:

- *informational* — new product, size/flavour; available from XYZ; shop now open 24 hours; sale starts Monday
- *interest generating* — free tasting; test drive; pick up a brochure; send off this coupon
- *educational* — milk is high in calcium; this product promotes root growth, children's health
- *reassuring* — the most effective pesticide on the market; environmentally-friendly; kind to animals
- *persuasive* — try/buy; exotic, smart, new improved; value for money; status-enhancing.

In the limited exposure available for most promotions only one message should generally be conveyed (the unique selling proposition), and the medium should be chosen for its ability to convey that message. Secondary messages (price, new improved product, solves a particular problem) may be included provided they do not obscure the main message or overload the observer. Secondary messages can be conveyed and the overall information content of the message can be significantly increased by associations of ideas and visual and auditory clues which are attention-catching or memorable. It is the function of good design, particularly, to maximise the ability to convey multiple messages by combining stimuli and evoking associations of ideas (for example, by humorous references, familiar characters, a brand-related musical jingle; in print media by print style, page layout, colour, paper quality, page/programme location etc.).

The content of a communication may also be enhanced by appealing to customer characteristics identified by the market research and appropriate to different media. For example, 'analyticals' may be targeted with a strictly rational, high-information presentation which stresses functional attributes or financial benefits. Other groups may respond more to an emotional appeal which conveys little information about the product but gains attention and interest by (say) nostalgia, humour, excitement, mystery. A provocative, argumentative or challenging approach may command attention. A pertinent comparison may sharpen the message: for example, between the real world and how it could be if the product were purchased (more convenient, easier to prepare, safer, tastier). Comparison with a competitor product is also a common technique (bigger, better, healthier than Brand X), but such comparisons must be accurate, and some regulatory authorities do not like them.

MEDIUM

A combination of media will normally be used to achieve different objectives. (Table 11.2 illustrates a media mix appropriate to a producer of a food product with a strong 'healthy' image.) Paid advertising and various forms of paid

- Generic TV and press advertising organised by a producer group/health food manufacturer or trade association in collaboration with government, to promote a healthier diet.

- Attendance at food trade exhibition to make contact with potential distributors who taste the product, take away literature on product and company.

- Firm follows up with full product presentation to distributor, explaining production process, inputs, quality control, merchandising support; negotiates terms, obtains listing.

- Firm uses this order as basis for publicity in trade press by inviting journalist to visit premises, write feature article, and promote company name to other distributors.

- Consumer advertising via press, local TV and radio, in collaboration with retailer to coincide with product launch/availability; accompanied by celebrity visit, posters/recipe leaflets, product tasting in shopping precinct outside retail outlet, doorstep drops of promotional flyers with money-off coupons, sponsorship of local school sports day/charity event etc.

Table 11.2 Representative media mix for a food product with a strong healthy image

and unpaid publicity are the principal means of attracting initial attention and stimulating interest, and in reviving fading interest and generating repeat sales (reminding and reassuring). To stimulate desire and actual purchase, sales promotions and personal selling are used, supported by advertising to reinforce the message.

Advertising

Advertising is defined as any form of paid, non-personal (mass) communication, which includes direct and direct-response marketing (catalogues, mail shots etc.) as well as TV and press advertising. *Above-the-line* advertising incurs a direct cost (design and execution fees, TV/radio transmission charges, press/poster space costs etc.); *below-the-line* advertising includes all other forms of mass communication for which no fee is payable (sales promotions, press announcements, PR, point of sale display etc.). Above-the-line advertising can be very expensive, and although it may achieve good exposure, there is no certainty that this will achieve sales. The advertising industry and most firms would not admit it, but the effectiveness of mass advertising is questioned except as a means of maintaining brand awareness: consequently, above-the-line activity always needs supporting by below-the-line promotion, even where a firm has a large advertising spend.

Small businesses commonly undertake above-the-line advertising at a local level, or in association with a regional or national promotional effort (for example, regional tourist boards and marketing groups), with a professional input in the design and execution of promotional material. For a larger scale promotion which requires higher expenditure, the cost of an advertising agency must be recognised as necessary to ensure that the right message is communicated via the right medium. The buying of advertising space/time is invariably easier and more effective through an agency.

As well as the medium, the timing, exact location and presentational format have to be decided, and the agent who deals regularly with media has the necessary knowledge to determine where value for money will be achieved. There are enormous differences in cost, for example, between a front cover advertisement in a local paper, a special interest newspaper or magazine, or a top-selling women's magazine, and the cost has to be assessed in relation to their relative effectiveness in reaching the target customer. The media which gain income from advertising (newspaper/magazine publishers, TV and radio channels etc.) provide information which allows the individual client to compare this, but this is certainly an area where a specialist input repays the investment.

The selection of the right medium will depend partly on the character of the message (informative, persuasive, reminding), but the objective is always to gain the highest level of exposure to the target audience in the most cost-effective manner, which invariably requires insider knowledge. (What is intended by exposure needs to be clarified, for it may mean number of potential customers exposed to an advertisement, or frequency of exposure.)

Direct and direct response marketing currently represent the fastest-growing advertising medium, responding to the increased availability of information technology and customer information generated from a number of sources including EPOS data. This information may be purchased from specialist firms and data exchange systems which provide this service to advertisers, and it allows customers to be targeted with much more product information than an advertisement allows. In that sense it is potentially more cost-effective, though the customer may see it as junk mail.

At the small business level direct response marketing is a potentially valuable form of advertising at a relatively low cost. For example, a farm shop or processor selling through local outlets can direct-mail local residents with money-off coupons for a product: a classic combination of advertising and sales promotion. Past customers of an accommodation or leisure enterprise can be targeted with promotional material advertising new facilities, a special event or a discount offer.

Trade fairs (including international fairs where export markets may be targeted) are another particularly valuable promotional opportunity for small businesses, because they provide direct contact with potential customers, at a cost which is often subsidised by government or development agencies. High-quality professional design services will generally be available which an individual business could rarely afford, and the promotional value may be enhanced by regional or national 'corporate imaging'. The opportunity to establish contacts with other producers and government agencies may also identify new opportunities for joint marketing and further funding assistance. Similarly, the opportunity to compare and learn from competitor techniques and promotional activities is an invaluable by-product of attendance at trade fairs. Exhibitions and fairs open to the public exploit the consumer's readiness to combine shopping with entertainment, and allow direct product sales, distribution of promotional material, and potentially high exposure.

Where small can look big

The International Food and Drink Exhibition (IFE) plays a key role in the development of specialist food companies ... The visitor profile is spread across the spectrum of buying power within the industry - wholesalers, retailers, caterers, both from the UK and abroad ... Niche companies need the kind of exposure IFE offers, as they do not have the marketing budgets or the manpower to gain the attention of major retailers looking for specialities. IFE provides that opportunity.

One young company exhibiting this year produces a range of pickles and chutneys hand made in a farm kitchen ... 'For a young company, IFE is the ideal place to launch a new product because of the broad range of buyers visiting the exhibition,' comments the producer, John Chenery. 'It's a good place to start your new product development too, just by listening to the buyers and then going away and responding to their needs.'

The Grocer, 4 January 1997

Box 11.3

Publicity

Publicity is defined as any form of non-paid, non-personal communication (posters, press releases etc.), which is generally understood to include public relations (PR) and media relations. In fact, all three are increasingly paid for, but they are distinct from advertising in being directed towards the establishment of a corporate image, whereas advertising generally has short-run objectives (though it contributes to the corporate image via brand awareness).

The increasingly high cost of advertising means that every opportunity for non-paid publicity should be maximised, and small firms have proved themselves very adept at exploiting opportunities which include press releases, press conferences at a product launch or opening of premises; news stories for local press and TV/radio; customer newsletters; open days etc. In a local area it is relatively easy to gain free publicity: local newspapers and radio and TV channels are always in need of good copy, and a local business doing well is good news. Since information is commonly syndicated between local and national news media, this can lead to national coverage. A direct approach to a national newspaper or magazine may also be successful in gaining wide coverage and a prestige endorsement. Managers must beware of the increasing use of 'advertorial' promotions, however, where the business featured in a magazine article is asked to pay for the exposure. Advertorial copy has been used to great effect by some diversified farm businesses to promote an accommodation or a food processing enterprise, and is certainly much less expensive than 'straight' advertising. It is nevertheless important to recognise that an apparently flattering approach for copy by a national magazine is not always cost-free.

An important part of publicity and PR is to demonstrate public responsibility, especially in food and family-related product sectors (environmental care, animal welfare, safety, hygiene). Sponsorship of local and national charities, sports and institutions provides a relatively inexpensive way of achieving this objective and reinforcing other aspects of a promotional campaign. PR is also inseparably related to product/business identification, which is communicated not just through a brand name, but via packaging, livery and other visual symbols associated with retail or wholesale outlets.

Sales promotions

Sales promotions are direct inducements to purchase which are normally price-linked, and related to new product attributes which add value to the standard product ('now with real butter', 'microwavable'; 'completely refurbished', 'new swimming pool'). At the consumer level sales promotions are familiar in the form of money-off coupons, point of sale display, demonstrations, in-store sampling, and on-pack competitions to encourage purchase. Similar inducements at distributor level include buying allowances, straight discounts, free merchandise, merchandising or promotional support (window dressing, in-store sampling), and even distributor competitions.

Sales promotions are generally agreed to be effective in generating short-run sales increases at both the consumer and the distributor level, though they may only transfer demand from one period to another and not increase total sales. They are an important part of the drive for competitive advantage, however, because they actively seek to win sales from the competition. They are also widely used for production-management purposes: for example, to smooth seasonal peaks and troughs in demand and supply; to shift stocks, or to stimulate demand in advance of new production launch or capacity expansion.

Sales promotions are usually supported by short-run advertising or unpaid publicity, which often borrow celebrity or media endorsement of products: for example, they can be timed to coincide with the launch of a cookery book or TV series, or holiday programmes on a region or themed activity. Discounts on accommodation can be offered to coincide with a local event which will attract visitors to an area. Small businesses also capitalise on large firms' advertising to promote their own competitor product ('piggy-back' PR). Forthcoming promotions are usually announced in advance in the trade press to maximise exposure, and this provides the opportunity to gain from their expenditure and effort.

COST AND BUDGET

Most forms of communication have high fixed (set-up) costs and variable costs for execution (air time, media space). Many small firms see this as unaffordable, given the relatively small scale of their business, and there is certainly no point in generating business which cannot be supplied or involves excessive management effort, or entails a level of development which was not sought and may be counter-productive to other business objectives.

One solution is collaborative promotion, for which varying levels of public finance exist in different countries and business sectors. The commonest method in consumer markets is a jointly-funded project between producer and retailer. The fixed costs are usually borne by the producer, who makes advertising copy, samples, and merchandising material available to retailers without charge, provided the retailer uses them to promote the product. The method is characteristic of generic advertising and group branding of agricultural products, which is funded by a levy on all producers because the object here is not to expand one producer's product at the expense of other producers, but to expand demand for the product in competition with other product sectors.

Setting a budget for the promotion programme and its individual components is very difficult, but in practice the critical issue is what is affordable: 50 per cent of businesses identify this as their budgeting approach. A sum is arrived at on the basis of fairly general financial criteria (what the corporate budget indicates is available), and a target level of expenditure is determined without reference to the strictly promotional needs of the campaign. The campaign then has to be tailored to fall within the budget.

A second approach relates the budget to expected sales, usually by reference to industry norms: for example, 10 per cent of sales revenue. A third (common) approach is the *objective-task method*, which determines the cost of performing a task to satisfy a desired objective. If the objective is, say, to inform X per cent of the population that a new product exists, the cost of achieving the necessary media coverage is determined. This cost is then compared with the available money to determine affordability. If it can be afforded the programme goes ahead; if not, the objectives and methods are revised until a budget is arrived at which meets reasonable management objectives at an affordable cost. The data used in this approach is built up from the business's own experience, but most agencies have data banks which will achieve the same objective (for example, TV viewing and radio listening figures, press readership, view figures from poster sites).

The objective-task method is used in budgeting for non-advertising communication activities, PR, exhibitions and leaflets, drawing on data from each experience. For instance, if attendance at a trade fair costs £5,000, but the business generated from last year's attendance was £50,000, the expenditure was worthwhile.

EXECUTION AND EFFECTIVENESS

Execution involves design and realisation of promotional material and communication via the right media, which requires time and skills few managers possess. Consequently, except on a very small scale of business, execution is normally delegated to a professional agent who will consult with management at every stage. Large companies and marketing groups generally put an idea to an agency (for a competition-based promotion, for example, or new brand livery) and review progress at regular stages to ensure that their brief is being met. For a small business, the local newspaper's advertising department and local television and radio companies will generally provide an acceptable design service.

Realisation may entail nothing more than colour printing or poster making. Increasingly, it includes video presentations and advertisements which, even at the small business level, involve actors, musicians and special effects, which add substantially to the associated time and cost. This must be borne in mind in evaluating alternative messages and media, and in assessing the need for a professional input and the affordability of a given promotion mix.

The effectiveness of promotion is measured by the extent to which it achieves the stated communication objectives. Standard market research techniques are used to measure customer awareness (consumer survey, TV/radio/media audience figures). Media interest can be shown by citations; clippings and press auditing services exist to provide this information, which is normally provided as part of the promotion agency service, or by an independent assessor. The type of information required, based on the AIDA model, is shown in Table 11.3.

Managers naturally wish to measure effectiveness in terms of sales increases, and research companies can provide models which allow this to be done. The output needs careful interpretation, however, because in the real world there is no straight-line causality between advertising expenditure and sales, since changes in the marketing environment and competitor activity during the promotion can distort sales. For instance, an increase in mortgage interest rates will reduce disposable income and depress sales. Although an advertising campaign may have been very effective in generating interest, it may therefore fail to generate sales. In this case, consumer belief in the product and wish to purchase would be acceptable measures of effectiveness, suggesting that sales will be generated when disposable incomes return to their former level, and indicating the benefit of a short-run price promotion to sustain the brand.

The effectiveness of publicity and reported PR is normally measured by converting citation figures (seconds of TV/radio exposure, column inches of text) into an estimate of equivalent advertising expenditure: a service which is again provided by citation agencies. This estimate can be compared with direct advertising expenditure relative to impact as measured above, to establish the cost-effectiveness of the promotion. The effectiveness of attendance at trade fairs and exhibitions is normally measured at the event by the number of enquiries received and brochures/leaflets taken; after the event it is measured by number of enquiries received and resulting orders taken. Trade fair organisers also collect this information and make it available to demonstrate the effectiveness of the event.

ATTENTION	Recall of:	Product identity, brand logo, point of sale material, advertisement

INTEREST	Content recall:	Advertisement, point of sale material, trade stand
		Take-up of coupons, free samples, leaflets, requests for information

DESIRE	Attitude/belief change Intention to purchase Repeat purchase Request for sales visit, technical presentation

ACTION	Sales made Habitual behaviour Brand loyalty

Table 11.3 The AIDA model applied to effectiveness evaluation

CONCLUSION

Promotion is an integral part of market segmentation and product positioning. The right balance of promotional methods depends on the business, but the same principles apply whether it is an owner-managed business, a large company, or a group involved in joint promotion. The effectiveness of promotion depends on the development of a clear, unambiguous message delivered via the right medium to the targeted customer, and its effectiveness is measurable not just in short-run sales increases, but in a long-run reputation which underpins a sustainable business.

CHAPTER 12

Group marketing

The nature of the agrifood sector is such that no producer, however large, is likely to be able to supply the volume demands of the market, or to achieve any degree of security or bargaining strength in a market dominated by a few very large buyers. We have also seen that many added value activities, product branding and effective promotion are difficult for the lone producer to achieve. The only realistic way in which many farmers can improve their marketing is consequently through joint ventures with other producers, which provides the necessary scale of operation, the organisation and the professional marketing skills.

Group marketing is already a strong feature of agricultural marketing. Many of the examples of effective marketing cited in this book are the result of horizontal coordination between producers who have combined their efforts to achieve negotiating objectives or undertake some processing or trading activity. The commonest form of association in most countries is the voluntary producer cooperative; in the UK and most Commonwealth countries it was formerly the commodity marketing board – effectively a compulsory cooperative, introduced in the 1930s because voluntary cooperation had failed. In the 1990s marketing boards were abolished in many countries (though they remain strong in some), providing the stimulus for the formation of new farmer-controlled businesses, some of which are incorporated under cooperative law and others organised as public limited companies.

All countries also have many different kinds of producer-linked bodies which undertake promotional or market development activities. Most are funded by a levy on produce sold or some other output or land-related formula, and they have different remits. Some are primarily R&D organisations with limited promotional or market development activity: for example, the UK Home Grown Cereals Authority. Others, such as the UK Meat & Livestock Commission, are very active in all marketing areas. Some promotional bodies, usually export-oriented, are funded partly by government plus income from services rendered: for example, Food From Britain, SOPEXA(France), CMA(Germany).

Many farmers are nevertheless deeply suspicious of 'cooperation', and argue from a few well-publicised past failures that all joint ventures are ineffective – though they do not argue from the failure of private firms that all such firms are ineffective. At the other extreme, some farmers remain rigidly attached to cooperative principles in the face of clear evidence that organisational change is necessary to keep cooperatives in business and achieve their members' objectives.

The truth is that the best run marketing groups are indistinguishable in their management and performance from comparable private firms. Some of the largest businesses in the European agrifood sector are producer cooperatives whose success proves that the cooperative form of organisation is not inherently weak. There is also ample evidence that past failures were due to poor member commitment, amateur management, and a narrow interpretation of the principles of cooperation.

The object of this chapter is to clarify the organisational (management) requirements which will allow joint ventures to achieve comparable efficiency and competitiveness with other firms. The account focuses on the particular problems associated with the cooperative as such because this is still the commonest form of incorporation in most countries, and as experience in the 1990s proved, cooperatives are perfectly capable of adapting to meet the requirements of a modern marketing system without compromising the principle of producer control which is their unique advantage.

Milk Marque calls for world class UK plants

The Chief Executive of farmers' coop Milk Marque said the UK will have to become an exporter if it is not to have an excess of unsold milk. 'There is a need for investment in world class plants within the UK, focusing on substituting for imports, finding new export markets and meeting and creating new customer demand.' He also indicated there was a possibility that Milk Marque might become involved in processing - but not before quota relaxation - for example by entering into joint ventures. If the rest of the industry did not respond to the challenge of the additional raw material then farmer-owned organisations would have to ensure there were adequate facilities available.

The Grocer, 1 February 1997

Box 12.1

OBJECTIVES OF GROUP MARKETING

The objectives of group marketing have been referred to throughout this book, but they are summarised below.

- *To increase sale prices*, on the premise that organised selling achieves a scale of operation which improves bargaining strength, and makes it economic to add services when selling large volumes of product.

- *To stabilise producer prices and/or revenue* by improved continuity and stability in product markets, and by appropriate pricing systems (for example, pool pricing).

- *To provide product branding* and product assurance, and promote the branded product.

- *To create or maintain a market for products* by deploying more marketing effort and, again, through economies of scale. No single producer, however large, is likely to be able to supply the constant, large deliveries required by retailer multiples; few could take on the additional responsibility of an export market or a credible advertising campaign. Collaboration may also be necessary to establish a market for an innovative product or where, for historical or locational reasons, there is none: a marketing organisation to distribute new crops, organic output; auction marts in remote areas, electronic auctions. Market development work and generic promotion are also required which may not be justifiable without cooperation.

- *To add value to products*: for example, livestock slaughter and meat cutting and packing; butter and cheese making; sorting, grading, packing and part-preparation of farm crops. To achieve least-cost production and exploit the scale economies common to processing activities, regular large supplies of raw product are necessary which may only be achievable through group marketing. Storage, transportation or grading may be operationally achievable and financially viable only for a producer group (for example, the sorting and grading of fresh produce, where the highest prices for fruit and vegetables are available only for uniform samples achieved by assembling and sorting large amounts of product). Poorer quality product can also be processed into a useable form provided there is sufficient quantity.

Cornishmen marching on

Cornish growers have begun to successfully convince major supermarkets like Tesco of the value of promoting a regional identity. Prepacks of cauliflower and spring greens carrying both the new Cornish King logo and the grower/packers' identity were taking pride of place on the Cornish stand at IFE. Flowers are also in the range, which will be extended to potatoes.
Roger Whilding, executive responsible for the campaign, funded through SW Horticulture 2000, said, 'Its acceptance represents a change in retail policy, giving growers an incentive to strive for better quality. About 20% of these crops are branded.' The scheme is backed by other disciplines. This season, Rocket, an early potato variety, will not be planted, and the logo was temporarily withdrawn during the cold weather to protect the image.

The Grocer, 15 February 1997

Box 12.2

To achieve these objectives farmers have formed two kinds of joint venture. *Operating groups* undertake processing and trading activities in the market alongside private firms, in the attempt to improve members' returns through competition and economies of scale. *Bargaining associations* negotiate with buyers on members' behalf, in the attempt to use collective strength to increase member returns.

OPERATING GROUPS

The easiest gains to achieve are those which result from the exploitation of economies of scale. This was the rationale of the earliest input supply cooperatives, and has now been developed by marketing groups. In many processing and manufacturing activities there are considerable cost reductions to be achieved by large-scale operations, principally by sharing the fixed costs of plant and management over larger throughput. Given the comparatively small size of most farm output compared with the needs of even a modest processing plant, some form of coordinated supply can yield obvious cost savings. Increased volume allows greater continuity of supply and reduces the risk of shortage caused by unforeseeable events, and allows a better service to be provided to customers who may be prepared to pay higher prices for continuity and consistency.

Scale economies also arise in promotional activity and media advertising which the small business could never achieve or justify. Economies of scale arise in the purchase of space or air time, but more importantly in the ability to spread the cost over different products if a brand is used. The value of the promotion is also increased where a single producer brand is used. This rationale underpins the activities of many organisations in the agricultural sector which undertake generic or collective advertising on behalf of a whole sector, benefiting from the increased value for money.

The achievement of higher prices in a market dominated by a few very large buyers is much more difficult. The motivation of many producers who form their own marketing organisation is the wish to increase competition in the market in order to transfer to themselves the 'supra-normal profits' which they believe the downstream sector makes at their expense. In theory the strategy is reasonable, but in practice groups find it difficult to achieve. Firstly, if there are no supra-normal profits, no gain can be achieved. Secondly, any group which succeeded in making the total market more competitive would benefit all producers, not just its members, while only members bore the organisational costs.

The process is easy to illustrate. Assume there is one buyer in an area who pays lower prices than the market will bear, and a producer group enters the market which pays higher prices to members. Other producers will join and sell to the group to benefit from the higher prices. In principle all rational producers will join, and the private buyer will be unable to obtain supplies. To stay in business he is forced to raise his price, at least to a level competitive with the group. All farmers in the area have therefore benefited at the expense of his 'supra-normal profits'. However, group members will have incurred costs in establishing the cooperative which non-members have not incurred, so members need revenue which reflects this investment. If the capital they subscribed to the group is remunerated by an interest payment the prices paid by both buyers can be equal; if this is not so, the group price must exceed the private trader's price in order to remunerate the capital invested.

The size of the revenue gain by members will also depend on organisational costs, and only if these do not exceed those of other competitor firms will members receive the benefit of competition. A marketing group must therefore organise itself to optimise operational efficiency and minimise the spill-over benefits to non-members.

BARGAINING GROUPS

Two kinds of bargaining position may be distinguished: distributive and integrative bargaining. The terms 'opponent pain' and 'opponent gain' suggest the meaning: in distributive bargaining, force is employed to redistribute the cake between the negotiating parties; in integrative bargaining, the size of the cake is increased so that both parties may have a larger slice.

Distributive bargaining

The obvious example of distributive bargaining is the industrial strike: withdrawal of the labour supply to force up price. Withdrawal of supply has been attempted in agriculture, but has never achieved its objective because it depends on total solidarity of suppliers and control over substitutes (imports of the withdrawn commodity from another region or country or substitute products).

In the unlikely event that both conditions could be fulfilled and short-term supply control achieved, the strategy is likely to be ineffective in the long run. Assuming (as we may) that raw material prices are forced up above the long-run equilibrium, one of two outcomes will arise. If processors can pass on the price rise to consumers, demand for the processed product is likely to fall and long-run revenue may fall. If processors absorb the price rise, profits will fall and firms will almost certainly become less wasteful, adopt higher grading standards etc., and seek out substitutes for the commodity. Theoretically, therefore (and there are many practical examples to prove the theory), the attempt to use opponent pain to raise farmgate prices is likely in the long run to result in producer pain.

Integrative bargaining

Integrative bargaining is much more likely to succeed. In this case producers, individually or collectively, provide services which buyers would normally perform themselves, but which the producer can achieve at lower cost or to higher quality standards. Lower cost may result from better knowledge, or the opportunity for a group to exploit economies of scale not available to the individual producer or buyer. Groups may also obtain free services from members for which the buyer would have to pay: for example, administration commonly undertaken without remuneration; uncosted labour in the supply of logistic services (for example, procurement and assembly of livestock by producer groups). Groups may also contract as a whole for a certain tonnage of crop which is divided among its members, reducing the danger of crop failure and ensuing supply deficiencies. In this situation both parties gain, and producers have a bargaining advantage which improves their chance of negotiating a better deal.

Bargaining power of any kind depends on a degree of supply control, which it is suggested must be as high as 75 per cent of total supply to gain significant advantages. Though this may be true of distributive bargaining, control of a much smaller volume may be effective where both parties stand to benefit. The minimum degree of control required varies for each commodity, but it must be sufficient for the buyer to recognise it as significant, and to allow him to make worthwhile savings on his operations. To meet this condition in a country of relatively small production units, cooperation between producers may be the only way to achieve the necessary volume of supply.

COOPERATIVE PRINCIPLES

The objectives of collective marketing are identical whether or not an organisation is a cooperative or public company, but the cooperative has additional characteristics which may constitute strengths or, if badly managed, weaknesses which need to be addressed.

What distinguishes a cooperative from private firms is ownership. Unlike other companies which are capitalised and controlled by shareholders who usually have no other interest in their activity, a cooperative is defined as a business organised for and capitalised and controlled by its members, to provide services and/or goods on an at-cost basis. It is a 'mutual society' which by definition exists to secure the mutual advantage of its members, and in some circumstances it may therefore attract criticism as a cartel, which government may wish to regulate. This situation arose in the UK in 1994 when the England & Wales Milk Marketing Board was abolished and replaced by a cooperative – Milk Marque, whose membership in excess of 70 per cent of producers was (unsuccessfully) challenged by trade and regulatory authorities as constituting a 'misuse of dominant position'.

All cooperatives world-wide conform to constitutional principles first stated in 1846, which still affect the benefits members may obtain and the obligations likely to be incurred. They are outlined below with reference to the opportunities and problems which this creates for marketing cooperatives in the 1990s.

Voluntary membership

Members voluntarily choose to join and to accept the duties and obligations of membership. In most cases members sign an agreement to this effect, but this undertaking is often more honoured in the breach than in the acceptance, and many members obey the rules only so long as they see it to be in their personal or economic interest. This would cause problems for *any* commercial organisation, and seriously weakens a cooperative formed to negotiate on members' behalf. In a non-cooperative firm such breaches of contract (for such they are) would entail formal penalties or litigation, but even where legally binding membership agreements exist cooperatives are reluctant to enforce them for fear of alienating other members and undermining the business.

Open membership

Membership is open to all persons willing to obey the rules who fulfil certain rather basic criteria (only milk producers may join a milk cooperative); discrimination on almost any other grounds is inadmissible (sex, religion, ethnic origin). In practice most collective bodies employ some limited controls if only to manage the size of the membership, though such action is discouraged by regulatory agencies which may refuse registration as a cooperative, resulting in the loss of some benefits (for example, tax advantages, qualification for grant aid).

Democratic control

Democratic control of cooperative affairs was a fundamental and revolutionary principle of early cooperatives, based on one-member-one-vote (OMOV). This remains the normal operational basis of most cooperatives, but the principle may hinder business development by giving equal representation to large and minimal users. This is an important consideration for a prospective member, since the obligations of membership may have substantial financial implications for his business although his influence over these decisions is limited. One solution is *patronage-related* voting based on actual trading, which reflects members' real involvement in the business (perhaps to a maximum number of votes).

Members of cooperatives are also consulted and play a part in the decision-making process in a way that shareholders in ordinary companies do not. They decide the terms of trade, how members are treated, what obligations they have to undertake, and so on. In principle this should be a strength, and it is certainly what attracts many individuals to form marketing groups in which they can exercise a greater degree of control over the marketing of their output. Many other members of cooperatives regard this an unwelcome obligation, however, and seek no active role in the organisation, preferring to be anonymous patrons who benefit from its activity without accepting responsibility for its success.

Service at cost

The business objective of cooperative firms is defined not as profit for shareholders but as service to members. This is seen by critics as a source of commercial weakness because it is believed to discourage effective commercial management. By this principle, however, the cooperative is merely the agent of its members – an extension of their business, not a business in its own right (though it has the right to trade in its own name). All the cooperative's activities are therefore based on the principle of cost recovery, and it is strictly a non-profit organisation. Any surplus over cost is the members' own money, accumulated by inadvertently under-paying for product supplied.

Since it is clearly very difficult to run a business day-to-day on this basis, most cooperatives pay a market price competitive with other firms and return the surplus to members via a year-end payment calculated in relation to trade conducted (*patronage refund or bonus*). The primary difficulty this causes for the cooperative business is in building up sufficient reserves to finance necessary investment and business development.

Return to capital

Members of a cooperative traditionally receive only a limited return to capital invested, on the grounds that remuneration of share capital is not the object of the association. In some countries the view prevails that there should be no direct return at all, only remuneration via higher prices paid for members' products or lower prices charged for inputs. In other countries (including the UK and Ireland) it is accepted that capital should be rewarded directly, but at a level representative of the true cost of borrowing on the capital markets, and the maximum interest payable is limited by legislation. This distinguishes cooperatives from other companies funded by shareholders whose only return is the dividend they receive – a dividend paid out of profits after tax, whereas interest paid to cooperative members is shown as a cost.

Financing according to use

An important cooperative principle is that activity should be financed by member capital contributions related to trading. In practice members are generally required to buy a minimal number of shares to obtain an equity stake in the business. Unlike ordinary company shares these are not openly traded and normally change hands at par, whereas ordinary company shares fluctuate in value reflecting market conditions. There is consequently no incentive for members to subscribe more than the legal minimum, particularly since this also limits their individual liability (which is limited, as in a non-cooperative company, to shareholding).

The capital which can be raised via share holdings is thus frequently well below the amount needed to provide the facilities or services required, and additional finance is necessary. This may be sought by requiring patronage-related member contributions, and provided voting is unrelated to shareholding this does not affect the control of the business as it would in a non-cooperative company. In countries where member shareholdings are limited by law (the UK, for example) members are required to subscribe loan capital reflecting patronage.

This kind of financing is commonly employed by cooperatives formed to process members' output or manufacture goods, since they require large initial funds to establish the necessary facilities. Dairy cooperatives, for example, generally require the producer's financial contribution to reflect milk deliveries, which may be ensured by making regular deductions from the milk price (*product retains*) until the two are in balance. This is reasonable because it recognises that the cooperative is the marketing arm of the farm business, and if it did not exist members would have to finance their own activities to achieve commensurate benefits (added value, market outlet). However, this represents a cost to cooperative members which is not borne by producers for whom the services are provided by a non-cooperative firm. Where loans are provided, many cooperatives 'revolve' them over a predetermined period (usually 5–10 years) to ensure that they are provided as far as possible by members currently using services, and are in proportion to current usage.

COOPERATIVE ORGANISATION

In the trading conditions of the 1990s, inflexible adherence to organisational principles first laid down in 1846 is inappropriate to marketing cooperatives. It is also inconsistent with the objective of cooperation, which is to defend member interests, not the survival of the cooperative for its own sake. The principal problems have already been identified:

- voluntary association (leaving members free to leave and 'free-ride' as well as join)
- organisational structure
- management structure
- finance.

MEMBERSHIP COMMITMENT

It is implicit in the principle of democratic control that members should be fully committed and involved in the collective enterprise. Cooperatives cannot afford free-riders, who seek to benefit from membership when it suits them and trade outside when market prices are higher. Such behaviour is the main reason why opponent pain bargaining often does not work, since a cooperative seeking to raise its price needs control over supply. If prices rise, buyers will retaliate by seeking to persuade members to break ranks and sell a product (or a proportion) outside the system, and the incentive to do so increases as prices rise. If one or two producers break ranks and it is seen that no action is taken against them, the practice will spread until the organisation has no control of supply, and may collapse.

The voluntary aspect of cooperatives should therefore be restricted to a member's decision to join or not to join. Thereafter, the membership undertaking should be as binding as any other business agreement. The commercial reasons for this are obvious where a marketing organisation makes plans in anticipation of a certain scale of deliveries, and may enter into long-term contracts to this effect. If the produce is not then available sales will certainly be lost, and an individual member's pursuit of a short-term private gain may cost the organisation long-term business. This is serious for a trading cooperative, but may be catastrophic for one which has invested in packing or processing capacity, since a very small fluctuation in throughput can turn a small profit into a loss. Once a member has undertaken to supply a given level of product and meet delivery and quality standards, effective sanctions must therefore be applied to ensure compliance. This is normal in milk cooperatives, but widely resisted in other sectors.

ORGANISATIONAL STRUCTURE

Free-riding may be a problem because producers do not see a cooperative as an extension of their own business (their marketing division), but as just another buyer competing for their output. The reasons for this are complex, but the market conditions in which a cooperative was originally established are a major factor.

Some of the most successful cooperatives were established to market products for which no buyer existed in an area (or none at all, where new products are involved). Dairy cooperatives are the most obvious example, where the highly perishable nature of milk made it imperative to shift it quickly off the farm and transform it into storable products. Where such facilities did not exist farmers wishing to produce milk had to create them: hence the existence of strong dairy processing cooperatives in Denmark, Ireland, the Netherlands and New Zealand, which were established in the absence of competition from the private sector. Their success is consequently not due to a stronger commitment to cooperative principles in those countries: it reflects the fact that there was (and may still be) no alternative outlet for a perishable product. Where there is competition for a producer's output from the private sector and other cooperatives, attitudes to cooperation are likely to be more ambiguous.

Two conditions are necessary in these circumstances to achieve equivalent commitment: the cooperative must consistently perform at least as well as other buyers, and its organisation must make it plain that the cooperative is an extension of the member's own business. The former depends on effective professional management, the latter on effective member participation in decision-making – which is not incompatible with effective professional management.

Member involvement

Effective member involvement is difficult for organisations with very large memberships, which have more difficulty in establishing and maintaining a sense of participation than small, close-knit groups. Regularity of contact is relevant. Milk marketing organisations which provide almost daily contact establish a working relationship which is more difficult to achieve in livestock, grain or horticultural cooperatives, where the transactional contact may be very intense for a short period, then totally absent till the following year. Regularity of contact is insufficient, however, unless it also develops a sense of shared responsibility, which leads some critics to argue that only small, close-knit marketing groups will be effective. In a large organisation, the only practical means of achieving member participation is a cellular organisation, with local or commodity-specific groups represented on the main board. Imaginative and sustained efforts will also be needed to inform and involve members as fully as possible within the constraints of running a complex marketing business.

Since cooperatives in most countries have to satisfy certain structural conditions set out in legislation, it is difficult to establish from their constitution the degree of effective member involvement. Differences in their objectives and, more importantly, their operations are nonetheless identifiable which relate to their performance and the strength of member commitment.

COOPERATIVE BUSINESS OBJECTIVES

As in the individual business, the objectives of member-controlled businesses are diverse and often irreconcilable. In principle, different objectives should be easily reconcilable in a member-owned business where the owners and users of services are the same people, who are also involved in management. In practice this is not so, and it is often easier to accommodate different interests in shareholder-controlled firms, where it is clear that the objectives of management and the owners are likely to be in conflict.

For example, managers may seek control of a large business for prestige and even salary reasons, whereas shareholders require the firm to be profitable, and only when profitability is achieved are they concerned with size. The relationship between owners and managers in such firms is also at arms length and may be based on anonymity, which has certain management advantages. However, it also means that managers divorced from owners can pursue their own objectives with impunity, provided they deliver a level of profit/share value growth which prevents shareholders from selling shares to a firm that may remove the present management.

In a cooperative business the situation is much more complex, since each owner-user has objectives for his own business which are distinct from his objectives for the group (though the two are obviously related). Members are also involved in group management, and the resulting choices may involve fine judgements. For example, profit maximisation by the group may require a lower level of throughput than that necessary to maximise individual members' income. Multiple objectives may be sought which have different output implications (sales, profit, patronage refund, membership) as well as supply conditions (throughput level, membership).

Even when the objective is decided, together with the corresponding level of supply, there are two ways of regulating the latter, each of which may conflict with other objectives. One strategy would control the number of members, which conflicts with the cooperative principle of open membership; the other would regulate the deliveries of individual members, which raises problems of equitable distribution and creates the opportunity for free-riders to undermine the organisation.

In order to resolve such conflicts of objectives it is essential for a producer-controlled business to decide what kind of organisation it is. There are three possibilities.

- If it is a free-standing business, independent of and acting at arm's length from its owners (like a public company run by management under supervision by directors, but with minimal involvement of the owners) it will probably aim for size, subject to a minimum level of surplus. If it became over-large and very successful, thereby raising the value (though not the price) of members' shares, members could decide to restructure or sell off the business in order to realise the members' share value.

- If the organisation is designed to act as the marketing arm of members' businesses, the object would be to maximise the sale revenue to each member. This might mean selling at the highest possible price regardless of the volume of output a member chose to produce in order to optimise his own economic efficiency. The organisation's function would in other words be one of price-maximising disposal, and the driving force would be the efficiency of the production unit.

- The third solution which a cooperative business should be able to achieve, given the active involvement and solidarity of its membership, is joint profit maximisation for the organisation and members' own businesses. In theory it is possible to achieve an optimum level of turnover at which the loss of production efficiency on the farm is just compensated for by an exactly equal gain in efficiency by the cooperative – or, of course, vice versa. The level of turnover necessary to achieve this in a given situation can be determined, though it is doubtful whether it can be achieved with such precision in practice.

Members must therefore be clear about their objectives in establishing a jointly owned business, and clear about what it ought to seek to achieve once it is established, for this will govern how the business operates on a day-to-day basis. For example, does it seek to sell at the highest unit price and is it prepared to restrict supply to achieve this, or does it sell as much as members wish to sell at the highest price it can, without resorting to supply restriction? If it was established to operate a processing plant but the organisation cannot afford to pay the market price for the raw material, should the plant be closed and members' produce sold on the open market? Alternatively, should members be required to sell to the group to keep the plant operating and thereby contribute to the fixed costs, which would have to be paid entirely by the membership if it closed? If insufficient supplies are available from members, should the plant be empowered to buy from elsewhere? If cheaper supplies can be imported, should the group buy these for processing in its own plant and sell on members' produce at a higher price to a processor-competitor?

All these are decisions which management may have to make, which may cause conflicts between members as well as between members and management. As in the individual business a clear set of agreed objectives and operating principles is thus essential, and the very act of involving members in the establishment of these will lessen the potential for conflict and increase trust in the management procedures.

MANAGEMENT STRUCTURE

Provided an organisation has a clear brief and clear operating procedures, management problems resulting from inadequate relationships between producer-owners and executives should not arise. The *quality* of management in producer-controlled businesses is another question, and the fact that some farmer-controlled businesses in the past achieved poor management standards cannot be denied. However, this was largely due to the failure of owner-members to acknowledge the need for professional management. When professional management *was* appointed, they were also generally reluctant to pay salaries commensurate with the job. Moreover, there is still a tendency to interfere in day-to-day management, ignoring the fact that in any business a clear and accepted division of responsibility is necessary between owners, management board, and employed managers.

Difficulties will inevitably arise where owners are patrons and some are also elected directors, but the roles of the management board and employed management should be the same as in any other firm: the board decides policy which the management executes. As in most firms, however, managers will invariably have more specialist knowledge and expertise than board members (especially in technical areas like finance and marketing) where their input into policy should therefore not be discouraged. Non-executive directors may also be appointed whose outside experience can guide the board and provide an independent viewpoint. The practice is increasingly adopted by producer groups, who are conscious of the criticism that past problems experienced by farmer-controlled businesses could have been avoided if independent advice had been available and (more importantly) had been taken.

FINANCE

The other major weakness from which all forms of collective marketing organisations suffer is finance. It was indeed the perceived inability of the cooperative business to finance itself adequately which led many UK and Irish cooperatives in the early 1990s to change their status to a public limited company. With hindsight it is now acknowledged that a cooperative business is equally able to raise capital from the market provided it can demonstrate that it has effective professional management and active member commitment.

Cooperative finance derives from four sources:

- the very limited share capital provided by members
- some funds related to use (either loans or, outside the UK, as shares)
- reinvested surpluses if the members can be persuaded to accept this
- commercial loans from cooperative banks (where they exist) or normal banks. These are often substantial and have to be guaranteed by some form of collateral as well as being serviced, each of which presents major problems, especially in a growth situation.

Ideally any business should be funded as far as possible by equity risk capital remunerated out of profits/surplus, not in the form of interest-carrying loans. Non-cooperative businesses raise additional capital by means of rights issues of extra shares. Cooperatives have generally been unable to do this because members were unwilling or unable to provide extra capital either as lump-sum payments or through product retains. This led many cooperatives to seek alternative ways of providing equity capital for expansion. The most straightforward of these is to adopt the status of a public company in which shares are sold to non-farmer investors (employees and others), but control initially remains with farmer members who own in excess of 50 per cent of the subscribed capital.

In theory this situation can last indefinitely, but if large numbers of shares are issued or farmers sell their shares – for example, under income pressure, control will rapidly move into the hands of non-member, and possibly non-farming, interests. This can be prevented and additional capital obtained by creating holding companies which are producer cooperatives, and subsidiaries which are normal quoted companies in which anyone can become a shareholder. Such schemes were pioneered in Ireland by the Kerry group, and have been highly successful in generating substantial capital which has been used for business expansion, both through internal growth and diversification and by merger/take-over.

CONCLUSION

Collective marketing is almost certain to become an even stronger feature of agrifood marketing, because it is the only realistic way in which many producers can deliver the requirements of the market. The success of many marketing groups also proves that they can deliver a level of efficiency and competitiveness at least equivalent to that of private firms, without losing the unique benefit of producer control. Success nevertheless depends on active member commitment and appropriate management – whether the group be a cooperative in the strict sense or any other form of business.

CHAPTER 13

Export Marketing

Export marketing is often thought to be the preserve of large firms, and to require special management skills. In fact, many very small businesses trade successfully across national frontiers (Plate 20), and many firsthand buyers of farm output are engaged in export marketing, including producer groups (Plate 21). Export marketing is indeed a major objective of many producer groups, who seek to create new markets for agricultural output which is not saleable in home markets, or saleable only at a low price.

The same planning, organisational and control functions are also involved in international trading (Table 13.1), though greater attention must be given to environmental appraisal and consumer requirements and behaviour, which vary substantially from one country to another.

A major environmental factor in all export markets is government intervention, since governments commonly encourage exports because they earn foreign exchange, generate employment, and benefit the balance of trade. Much more business support is therefore available for export mar-

Ready to export?	*Terms of trade, logistics*
Check: • management expertise & commitment • product strengths & weaknesses • capacity for product/packaging modifications • production capacity relative to new sales • credit facilities to finance exports	Consider and arrange: • agent/representative/distributor • delivery terms • payment terms, credit period • currency and credit insurance • arrange pre-shipment finance where necessary • appoint freight forwarder/shipper • documentation
Identifying markets & making contact	*Keeping in touch*
Consider: • desk research/commissioned research • local/international competition • local technical standards • product/packaging/customer requirements • visiting the market • attending export seminars, trade shows/missions • meet potential buyers/agents/distributors	Consider: • how to maintain effective customer contact • how to promote the product • sales literature • follow-up trade shows • how to arrange after-sales service

Table 13.1 Export marketing planning

keting, from government departments and export organisations. Most food exporting countries have an agency dedicated entirely to this sector: for example, Food From Britain, SOPEXA, CMA. In the past many countries also subsidised agricultural export prices (EU butter, US grain, New Zealand dairy products) and/or inputs for export-targeted production.

Governments also intervene in international trade for political reasons: for example, the denial of exports to unfriendly countries, or positive discrimination in favour of countries where trade is considered beneficial to political stability. Political intervention in export marketing therefore has to be taken as given.

Most businesses enter the international market as exporters: they produce in one country (normally their home country) and export to other countries products which were initially designed for the home market. Most small businesses remain exporters, though with experience of the new market they modify the marketing mix to satisfy their target export customers. Some firms, including many farmer cooperatives, move on to manufacture export product(s) in the export market ('international marketing'), and the next step is to become a truly global company which manufactures components or whole products at the least-cost location for shipment around the world. This again includes some producer cooperatives: for example, MD, the Danish dairy cooperative.

The common objective is business growth which is not achievable in the home market. The attraction is that the product exists, production problems have been solved, and exporting may be a simple matter of diverting a proportion of output into another market. Capital costs are low, and if the initiative fails there is always the home market to fall back on. Additional costs are involved in research and market development, but they can be quite low if the exporter adopts a gradualist approach, and government aids to export development are a further incentive. However, export markets should not be seen as a dumping ground for surplus production, but as an opportunity for sustained business expansion which may spread the trading risk and make more effective use of the production base. The core home market should also not be neglected, because it is the fallback position if the export market is closed, as Box 13.1 illustrates.

MD stops Gaio sales

MD Foods has confirmed it is ending sales of Gaio yogurt in the UK due to poor demand and 'adverse publicity'... The yogurt contains the live bacterial culture Causido, and was marketed with a £4.5 million advertising spend since its launch ... on the basis that it lowered cholesterol levels. In March 1996 the Advertising Standards Authority upheld complaints that these claims were unproven ... MD Foods confirmed the product 'has been criticised ... and proved unpopular with consumers following seven months of adverse publicity.'

Gaio remains one of the most successful food products in Denmark (MD Foods' home market) where it has even out-sold butter.

The Grocer, 4 January 1997

Box 13.1

THE EXPORT PROCESS

The same stages are involved in the planning and evaluation of export marketing opportunities as in the domestic market, but more information is required since the target market is unfamiliar. Exhaustive analysis cannot be justified if the marketing opportunity is unlikely to be viable, so the country profiles produced by national export organisations provide a useful preliminary screening. Early contact with such organisations is advisable to identify the range of services available, which normally include:

- opportunity analysis
- market and marketing research data/services
- strategy planning advice
- assistance in distributor selection
- buyer contacts
- logistical support
- advice on advertising, promotion and trading policy.

The business objectives sought in considering an export market must be clarified, since they will affect the scale and extent of the resulting marketing commitment. Customers almost certainly exist for a small volume in a large market provided it reaches the right outlets, and the right export agent will achieve this. If a large volume is required, however, firsthand research and market development work may be required to sustain a long-run marketing programme. A multi-product business will rarely expect to export all its products, and almost certainly will not export them to the same markets. It may be appropriate to export only a luxury product to one country, or to well-targeted up-market niches in several countries.

Likely export markets will be suggested by business experience and capabilities, distance, politics, and cultural similarities. For most small European companies, for instance, the EU is not only the obvious common market, but the distances involved are manageable, logistics systems are known and may be relatively easily accessed; trading practices are fairly standard, and the socio-political and cultural

Economic

Costs, tax levels, import duties
Prosperity, growth prospects
Trading association (EU, NAFTA)
Exchange rate policy
Financial policies

Political

Political system, stability, ideology
Inter-country relationships
State control/private sector balance
Legal system, controls

Geographic

Location, size, climate, topography
Population distribution
Ethnic distribution

Socio-cultural

Development level
Literacy/education level
Religion/dominant philosophy
Language(s)
Social class structure, family pattern
Income, wealth, living standards
'Ethical', environmental concern

Technological

Public infrastructure
Transport
Communications, including IT
Household technology
Attitudes towards technology

Table 13.2 The export environment: data requirement

background is roughly comparable. (In trading blocs like the EU, sales to other members are strictly not exports but internal transfers; 'exports' are sales to third countries outside the bloc. In practice 'export' is commonly used, as it is here, to mean any sales across national boundaries.)

A shared language is a common factor narrowing the export market search: English-speaking countries world-wide invite trading associations because the common language and similar socio-legal structures over-ride distance. Expatriate or emigrant communities with similar product requirements to consumers in the home market are another viable target. Some small businesses with a very distinctive product may be able to trade world-wide without regard for language or culture: for instance, Scotch whisky, Stilton cheese, Parma ham.

The data required to evaluate potential target countries is essentially the same as in home markets (Table 13.2), but the political nature of the marketing environment gives priority to the trading policy of likely target countries.

TRADING POLICY

Although the world is committed by the 1994 GATT agreement to move towards global free trade in all commodities between all countries, the world is also characterised by trading blocs and national barriers which obstruct this objective. Some trading blocs practise free trade between member countries and external protectionism against third countries; the most advanced have created an internal single market which further encourages and assists internal trade while not dismantling external barriers. Consequently, although the world is nominally committed to free trade, protectionism is still significant and easily justifiable to domestic producers and consumers.

Within a trading bloc like the EU a product which can legally be sold in one market may in theory be sold in any other member country market. In practice trade impediments are regularly invoked – chiefly public health and safety considerations – which governments can easily justify to their own citizens. Such barriers are frequently illegal, and a major role of producer organisations is to exert pressure on their own government and trade regulatory authorities to ensure that rules are obeyed and a 'level playing field' exists in their target market. Most trade restrictions are intermittent, however, so they can only be reported and lobbied against once they have occurred, which is too late for the exporter who is already in a market. A business which relies significantly on export markets must therefore be aware of the risk, and the means which may be employed by countries seeking to restrict imports.

Import taxes

The most overt and internationally sanctioned import restriction method is the tariff or import duty: a tax levied by the state which restricts competition in the domestic market by raising the price of imports. These taxes may be levied as a percentage of the offer price; as a flat rate ($50/tonne); or as a rate which varies with the offer price, but is sufficient to raise the import price to a predetermined level high enough to disadvantage imports relative to home production (the basis of the EU CAP.) The tariff rates applied vary by product category and form: for example, many countries have lower rates on raw materials than on the processed product to prevent the export of processing jobs.

Import taxes are generally preferred because they are considered less distorting than other measures, for if they are applied at a fixed level they can be surmounted by production cost reductions in the exporting country; they may thereby encourage efficiency and allow the expression of international competition. This may lead to charges of dumping, however – sales at a price lower than economic production cost, achieved by government subsidy or simply by a seller disposing of surplus product at a very low price (perhaps based only on variable cost recovery). The USA has legislative provision for the imposition of such duties and has threatened their use against the EU on several occasions (for example, anti-dumping duties were imposed on EU cheese exports in the 1990s). The GATT agreement seeks to prevent anti-dumping legislation, but the procedures are again lengthy, and practical efforts by producer organisations and governments are more effective in deterring short-term dumpers.

Quantitative import control

Many countries seek to protect domestic producers or conserve foreign exchange holdings by imposing limits on the physical quantities that may be imported. These quotas may be imposed by the importing country or they may be voluntary restraint agreements (VRAs) agreed between the importing and exporting country (for example, agreements between New Zealand and the EU for butter and lamb, to avoid price collapse in the EU market which would disadvantage both parties). They may be bilateral agreements, or international agreements sanctioned by the world trade authorities: for instance, the multifibre agreement designed to protect the developed world from low price competition from developing countries. The 1990s saw efforts to reduce such import controls because they restrict competition and may thereby restrain economic development. The substitution of tariffs (*tariffication*) has begun which will undoubtedly assist trade in agrifood products, where import controls were particularly prevalent.

Product-type conditions

International trade has commonly been restricted – even within free trade areas and single markets – by the imposition of 'type conditions' which seek to exclude products that do not conform in every detail to a domestic equivalent. These restrictions were very common in the former EEC (they were used, for instance, to restrict foreign beer imports into Germany), but the Cassis de Dijon judgement in 1979 declared it illegal to attempt to restrict the sale in any member state of products legally saleable in any other. (Whether consumers will buy the product is another question, but its importation cannot be restricted.) Type conditions continue to impose barriers to trade with third countries, so the acceptable product characteristics to fulfil these requirements must be identified and complied with.

Health and safety

Product-type conditions are often imposed for public health and safety reasons, which makes them particularly common in the agrifood sector. The regulation of international trade on these grounds needs no justification to the population of a country, whose government cannot afford to be seen to neglect public safety. Exporters must therefore abide by health and safety rules in any export market, and in the last resort this may mean that a product simply cannot be exported. For example, only products produced in prescribed conditions may be permitted, or certain procedures may be disallowed. Labelling regulations and advertising codes also differ between countries on health and safety grounds.

All countries impose phytosanitary and animal health conditions on imports of plants and animals and products derived from them, and such restrictions are generally absolute, allowing no exceptions. They may be imposed instantaneously and take immediate effect, in response to food or health scares which grip a population. Their impact on exporters is therefore immediate, and in the short term nothing can be done about them, even if they can be shown to be ill-founded. Market diversion may be possible, but such scares usually affect whole regions simultaneously and may even be world-wide, closing all potential markets. In the long term production may be adjustable to allow exports to re-start, but only at some cost (for example, livestock systems may exclude hormones and growth promoters banned in some European markets)

Efforts are continuously made to reduce the impact of health and safety issues on international trade by agreement of common standards for products, premises and personnel, uniform labelling rules, etc. Progress is very slow, however, and the use of health and safety claims to exclude unwelcome imports is a permanent threat.

Non-tariff barriers

Product-type and health and safety regulations are non-tariff barriers which happen to be important in the agrifood sector. Governments use many other criteria to restrict international market competition, and new ones are continually introduced. Political acceptability has often restricted trade with some countries (South Africa, Iran, Iraq) and promoted trade with others (Taiwan, some developing countries), and a country's environment and animal welfare record as well as child labour laws are now also invoked.

The feeling within a trading bloc that some members are not wholly committed to free trade may act to exclude its exports: for example, EU reaction against Eurosceptic views in UK politics. A 'buy-domestic' policy may also be imposed on public bodies (and must be distinguished from the natural inclination in many countries to buy domestic produce, which is certainly an obstacle to foreign food imports but is perfectly legal). Again, such a policy is easily justified to the domestic electorate as safeguarding jobs or spending taxpayers' money at home, and is therefore difficult to eliminate. Excise duties and other types of domestic taxation – for example, varying rates of VAT within the EU – may also restrict competitive imports (wine against beer) or restrict demand. These are particularly resistant to change, however, because even within a single market governments retain power over what is a major revenue-raising instrument.

Export restrictions

Although most countries wish to encourage exports, they are sometimes discouraged or prevented. Exports of a commodity in short supply may be halted to prevent the shortage falling on the domestic market and raising domestic prices. In 1995, for instance, a shortage of grain in the EU brought a cut in export subsidies to make exports less attractive than domestic sales and curtail the rise in animal feed prices, which would have led to higher milk and meat prices. Commodity exports may also be restricted to prevent short supply to domestic processors who add value and create employment in an economy. This is particularly common in developing countries, but is an expedient which is also used by the most advanced economies: the USA, for example, has restricted soya bean exports to maintain supply to domestic processors.

Such restrictions may be imposed as an export tax (often levied as a source of revenue rather than to restrict trade), by export quotas or by export licences which may be allocated, sold or given as a form of patronage to the potential exporter. Developed countries occasionally use export taxes over a limited period, but export licences are widely used to control trade in strategically sensitive materials/manufactures, or trade to unfriendly countries.

COUNTRY PROFILE

The starting point for the analysis of export opportunities is a country profile, often available from national export organisations, which should provide the following information:

- physical and major socio-economic-cultural features of the target market country (mountainous, subject to climatic extremes, religion, language etc.)
- level of economic development and rate of growth
- socio-political structure, form of government
- demographic features (density, location, age/family structure)
- culture, traditions, lifestyles
- logistical and commercial infrastructure.

Physical and broad socio-economic data are easily obtained (though economic information may be dated and is not necessarily reliable), and this alone will often exclude many potential target markets. The other factors require more detailed comment.

Economic development

A market for almost any product, albeit a small one, can probably be established in any developed country, given its relatively high level of disposable incomes. This is not normally true of developing economies (including former Communist countries), where foreign exchange tends to be short and non-essential imports are restricted. However, the process of development is continuous and in some countries has been extremely rapid, particularly in newly industrialising countries (NICs), where opportunities to export specialist products have expanded along with income. Rather than simple income considerations, more likely constraints are therefore literacy and educational level, local technology and access to it (including basic household technology and reliable pure water supplies which cannot be taken for granted). Also critical are the nutritional level and expectation of the population which affect attitudes to food products: for example, developing-country consumers will be much less receptive to organic, reduced-fat, additive-free, and environmentally-friendly claims.

On the other hand, the same circumstances may present opportunities for low-tech products, and for products which are at the mature stage of the life cycle in the home market (for instance, canned and dried foods, including nutritionally balanced whole meals). As access to refrigeration increases, exports of frozen foods are expanding rapidly. Within the poorest countries export opportunities also exist to supply the needs of wealthy, often western-educated consumers who seek western added-value products. Though normally very few in number, they share the same wants as expatriate communities resident in these countries, so the two may often be served together. They also have an important demonstration effect in introducing products to consumers with rising incomes, and their political power opens many doors.

Socio-political structure, form of government

The political and social structure of a country affect both consumption and market access. The principal consideration is the target country's trade policy and its willingness and ability to pay in convertible currency. Many other potential obstacles arise from different attitudes to private trade (as opposed to state purchase), attitudes to entrepreneurs (indigenous as well as foreign), and legal and commercial structures and trading practices.

All these may affect terms of trade, the sanctity of contracts, agents' fees, repatriation of profits etc.

Less formally, the ways of doing business differ widely, and exporters must recognise this in order to avoid transgressing the local code and losing the business. This is usually fairly evident in countries with obviously divergent historical and socio-political traditions, but exporters commonly underestimate the differences within a reasonably homogeneous society like Western Europe, North America or Australasia. Though countries within these areas share the same basic values and many common trading and legal arrangements, the differences between the USA and Canada, between New Zealand and Australia, and between France and the UK are significant hurdles.

Demographic features

The size and population composition of a country are obvious determinants in any export evaluation, but as ever, careful interpretation of data is essential. An exporter wishing to sell large volumes of product might think a market of 10 million is more promising than one of 1 million, but gross statistics of this kind can be misleading. A country with 50 million inhabitants who do not eat lamb may seem a great opportunity, but the opportunity may be zero if there is some cultural barrier to consumption or they cannot afford to buy it. A large market with no established consumption of a product may therefore be less attractive than a smaller market where consumption is established: for example, per capita lamb consumption in Switzerland is higher than it is in Germany.

Population density and the presence of large population centres are relevant to the logistical evaluation and the ability to communicate directly with the relevant market segment(s). As in any other market, demographic factors like income, age structure, and ethnic mix will enter into the analysis. For an exporter of luxury products, for example, a country with a high proportion of affluent consumers is a more likely target, but this could be Germany, or the tiny sub-market of very affluent consumers in a developing country.

Culture, traditions, lifestyles

Some generic customer characteristics will not need detailed analysis. Religious belief is an obvious example; there is no point in exploring Muslim markets for pig-meat products. Less obviously, it is important to determine who purchases and who decides what is purchased: it may be servants or men rather than women, and the relative influence of children's pester power in India and the USA may be very different.

Lifestyles tend to be similar throughout a country, and if the exportable product does not match these the market as a whole may be disregarded in the short term by small businesses. In the long term lifestyles can be changed, and this process can be speeded up by marketing activities. Lifestyles everywhere are converging, fuelled by popular US-derived culture and information technology, with an accompanying convergence of purchasing patterns. In theory this makes it increasingly possible to sell anything anywhere, but cultural convergence provokes opposition from groups and governments, and some consumers and suppliers resist globalisation and actively seek non-conforming products. Many niche markets therefore exist which are ideal for an exporter, whose product is by definition different and possibly unique, and may be seen as superior because it is foreign. An exporter with such high value-added products may therefore charge high prices for small volumes.

Infrastructure

Globalisation of products is accompanied by a globalisation of infrastructure. Differences obviously remain between developed and developing economies, but these are being eroded. In developed and newly-industrialising countries, products are sold in supermarkets, distributed by truck, and accounts are kept on computer. Radio and television can be received everywhere since electricity is almost always available, and refrigerators and freezers are becoming universal. Commercial infrastructure is available and effective; banking and insurance services, market research, advertising and management consulting agencies all exist; world-wide communication by telex and the Internet is widely practised. In this respect international trade is therefore becoming no more difficult than domestic trade.

CUSTOMER SELECTION

Customer selection within a target country does not differ from that required for domestic markets. Manufacturer and resale markets should be investigated as well as consumer markets, using information publicly available from the government department or organisations responsible for exports. Local consumer research may have to be commissioned to identify the product characteristics necessary to supply an identified market segment. Marketing research will establish the distribution requirements necessary to get the product to the target customer, to determine appropriate promotional objectives and means, and a price satisfactory to the customer and sufficient to meet stated profitability targets.

MARKET ENTRY

There are three basic ways of entering a market:

- direct export to the customer (industrial buyer, wholesaler, retailer or consumer)
- via an export organisation based in the producer country
- via an importing organisation based in the market.

These are not mutually exclusive: all three may be used as required by a particular market and customer group, product type, and the distribution structure in the market served.

Direct export

Direct export from producer to customer has considerable advantages because it allows the establishment of partnerships and close collaboration which helps to solve other marketing problems. In countries where multiple retailer central distribution exists, this provides a very convenient solution to market logistics, since a single delivery achieves easy access to a central depot from which the retailer distributes the product over a wide area. The retailer may also organise promotion activities, determine the selling price, and may even design the production specification (own-label products).

These are valuable services, but they need careful management to avoid some potentially damaging effects. They also do not eliminate costs incurred in maintaining an export department/competence, and the documentation costs involved in customs clearance at both ends, foreign exchange trading etc. If exports are a significant part of the sales it may be worth this expense to establish direct links with customers, especially as these links become less expensive to maintain once they are established.

Export/import agent

Many exporters start by using an export or import agency or merchant, and this is invariably wise in countries where there are still very many small independent retailers and intermediary wholesalers. Exporters and importers may be merchants who take ownership of the product and are effectively wholesalers, or agents who trade on commission (Table 13.3). All may provide market research services and promotion advice, and serve as the market presence of a product.

Most international trade and marketing is handled through agents, though changes in industrial and distribution structures have led to more direct trade. A high proportion of perishable foodstuffs, flowers and plant materials is still traded on physical markets in which agents buy and sell on behalf of their principals (for example, Rungis, Alsmeer). Less perishable commodities may be traded over the international commodity exchanges. Export-import functions are sometimes combined, generally by very large international trading houses.

Wherever products are sold by another party, control of the marketing programme will be diluted or may be lost. This should be less true where agents are used since they act on the instructions of principals to whose programme they should adhere, or from whom they should seek new instructions if they

AGENTS	DISTRIBUTORS
Negotiate with customers and take an agreed commission on value of product ordered. The exporter retains title and responsibility for transportation and invoicing to the customer.	Purchase (and take title), and are therefore responsible for distribution, price setting, and invoicing.
Advantages	*Advantages*
• High level of control • Commissions usually lower than distributor mark-ups	• One order/payment/delivery • Full marketing service delegated • Marketing costs split exporter/distributor
Disadvantages	*Disadvantages*
• High level of control • Commissions usually lower than distributor mark-ups	• Possibly high mark-up • Marketing control lost • Uncertain commitment of distributor

Table 13.3 Exporting: agent or distributor

wish to vary it (for instance, by selling to a multiple retailer rather than independents). Control is always lost when sales are made to a merchant; for example, he may discount a product to increase sales, at the expense of status or prestige objectives in the producer's marketing programme. This may be damaging if the same product is exported to other countries via agents who adhere to the original marketing programme, since both the product image and the price premium may be at risk.

The small business will probably have to start with an agent selected in consultation with trade associations and government exporting agencies, who often make the first contact and assist in negotiating terms. The latter also provide access to international trade fairs and shows at which small exhibitors may introduce products to the marketplace, and representation in trade missions. Both activities are often subsidised by domestic governments, because the presence of small enterprises on national stands and in trade missions is welcomed for the product width and differentiation they add to the national offer.

As exporting gives way to global trading, private companies and farmers' cooperatives are also setting up subsidiaries in third countries who take delivery of products, possibly from production sources in different countries, and sell on the company's behalf. This solves all the problems of marketing control and may solve currency problems.

MARKETING PROGRAMME

The marketing mix for an export programme will have the same components as in any domestic market, modulated to the target market.

Product

Most exporters seek to export a product which is identical to that sold in the home market. This may be acceptable for a regional or strongly differentiated product (Scottish shortbread, Stilton cheese), but a strong promotional component will be necessary to stress its unique characteristics. Generally it is not wise, and it may not be possible, to export exactly the same product successfully. As Box 13.2 suggests, modifications may be required to inherent product characteristics (form, sweetness, colour, flavour) or presentation and packaging (pack size and shape, meat cuts). In other cases more substantial product modifications may be required for market acceptance, involving a complete product re-design which undermines the financial feasibility of supplying this market. This needs to be determined at the earliest possible stage of market screening, by consulting trade associations and export organisations.

Product attributes subject to legislative control need careful identification; for example, different countries have very dissimilar standards for permitted food additives, product labelling, trades descriptions, promotion, and the environmental impact of products and packaging. High standards frequently begin in one country and subsequently spread to others, and a business which rejects market entry into a high-standard country may subsequently find itself obliged to supply to this standard elsewhere. It therefore makes sense to seek to satisfy the highest standards in all potential markets, or at least have a planned programme of product development to meet this target.

Cultural clash over tea-time treats

Tea on the lawn ... is seen throughout Europe as thoroughly English ... UK biscuit and cake makers with money, image and the home market on their side should therefore have little difficulty in dominating the west European market ... But the French buy dry mini-cakes for 'le gouter'... and there are few equivalents to the British packed lunch ...

Strong national, historical and cultural differences bar the way to easy exportation of biscuits. The British eat round digestives and ginger nuts in plastic roll packs, but these are unacceptable to French hypermarkets since they roll off the shelving - buyers are accustomed to square biscuits in expensive cardboard packs ...'Shipping stuff around because it tastes good in one market does not necessarily guarantee success in another.'

The Grocer, 18 May 1996

Box 13.2

Product brand names may not be directly transferable to another country because they have adverse linguistic or cultural associations, or are already in use in the export market. This is not necessarily a disadvantage because many brands which are well-known in a home market may be totally unknown in the target export market: no prestige would therefore be transferred with the name, so the product is essentially a new one, with a new brand to establish. The new name must be chosen with specialist local advice, however, to avoid unfavourable connotations or replication of an existing brand.

Price

Pricing in export markets must take account of:

- additional production and delivery costs
- exchange rates and their fluctuations
- the prices charged by competing products in that market
- the profit expectations of importers
- the possibility of charging a 'foreign product' premium reflecting the exclusiveness, status or prestige of the product in that market.

The resulting calculations are difficult for global or multinational businesses, and almost impossible for small businesses for whom exporting is a marginal activity. Sometimes the problem does not arise; for instance, the purchaser of an own-label product decides the consumer price and negotiates the supply price with the producer, who can only decide whether or not to supply. It is almost as straightforward when selling to a merchant, since the supply price is negotiated, but the price which the buyer charges his customer is his decision. The exporter may suggest a wholesale and retail price as he does in the domestic market, but agreement may not be achievable if the exporter and the importer have different pricing and marketing objectives, or achievable only at the loss of some marketing control by the exporter.

For example, a producer wishing to adopt a price skimming policy for the introductory phase in the market will recommend a high price, but a buyer may decide to discount the product for promotional reasons in order to achieve high market share. Alternatively, a producer may wish to build market share quickly by low price penetration, but this may conflict with the wishes of a buyer or even the traditions of the country. In the domestic situation these potential conflicts would normally be resolved by discussion, and a coherent policy for both parties would emerge. This is more difficult to achieve in international marketing, though it is becoming more normal.

The area in which the exporter has unquestioned control over the supply price is through his own efficiency. One reason for entering export markets is to exploit competitive advantage which may be cost based: lower factor costs or more efficient factor utilisation. A business can also choose not to recover all its costs on exported products, the export price being based simply on the recovery of variable costs, leaving all the fixed costs to be carried by the domestic market. This is in fact very common, and almost inevitable when a market is entered, but it is not sustainable in the long term and can lead to charges of dumping and consequent retaliation.

Currency exchange rates make pricing more difficult, especially for small businesses which cannot afford to undertake the currency trading activities of large global companies. In single currency areas or countries with fixed exchange rates the financial risks of international trade are much reduced, which is why exporting firms are interested in fixed exchange rates, and EU firms are keen on monetary union. Under conditions of floating exchange rates (albeit within limited bands) the foreign exchange risks of exporting can be quite high, when currency fluctuations of 10–15 per cent in a day as a result of speculative pressure are not unusual. Small businesses therefore have to build allowances for such fluctuations into their pricing, and many try to sell to merchants in their domestic currency, transferring the management of the risk to the buyer. If this is not possible, foreign exchange exposure may be reduced by using the 'treasury' services offered by commercial banks.

Distribution

Physical distribution of products to export markets is very market- and product-specific, and must be researched in every market. Within a relatively homogeneous geographical and trading area like the EU or North America, distribution will largely be a matter of identifying and accessing existing transportation and logistical support systems. There are broadly five options: road, rail, sea freight, air cargo, and courier services, the choice and combination of which depend on availability and cost, and factors like product perishability, value, weight:value ratio. If an export agent is used, the distribution system will normally be determined by the agent, and it is simply a matter of delivering as required. If direct exporting is involved, export organisations will provide advice on available means and the best combination to achieve efficiency, speed, and economy.

The documentation surrounding export trade can be formidable, and may be a major factor deterring would-be exporters (not unreasonably, in some markets). Export organisations and government departments are particularly useful in this area, advising on obligatory control procedures, the acquisition of health certificates, certificates of origin, export licences etc., and they provide language services for invoicing, shipping instructions, and instantaneous interpretation.

Promotion

Most exporters necessarily become involved in both pull and push promotional activity since consumer choice and behaviour must be influenced and credibility established with the distributor as a reliable supplier. The task is substantially the same as it is in domestic markets, but much more aid is available to small businesses wishing to export.

Why did the chicken cross the Channel?

For Harrison's, a successful family-owned British poultry company, a niche export market beckoned for their poussin. They had chosen to compete in a market traditionally a preserve of continental companies, yet they had first to compete with the channel. To succeed in this market, fresh, daily deliveries were essential.

With Food From Britain (FFB) experts, familiar with these markets, a feasibility study was undertaken. FFB contacted likely retailers and arranged for them to visit Harrison's UK factory. The response indicated a strong case for exporting and concluded in a deal with Delhaize, a major Belgian retailer with a listing in 110 stores. Harrison's beat off strong French competition and, together with Dockspeed and FFB, had no problem delivering.

After two years, exports account for 15% of total turnover - and are still rising. Harrison's flexibility and response to retailers' needs have opened up new listings for value-added poussins and private label ventures, including consignments to Denmark and a test market in Hong Kong.

Food From Britain, Exporting For Future Growth

Box 13.3

As in the domestic market the small business is not well placed to undertake large-scale consumer promotion, but this will generally be undertaken by the distributor, who may require a promotion contribution from the supplier (funds and supporting material). Alternatively, it may be organised on behalf of producers by a national export organisation in association with local distributors. The same organisations have a remit to establish the credibility of new suppliers, effectively by acting as guarantors that products will meet production and delivery specifications. This involves promotional activity, especially exhibits at international trade fairs, and the organisation of outward and inward trade missions (suppliers➔buyers, buyers➔suppliers).

Legal requirements with regard to advertising and promotion vary significantly from one country to another, and cultural standards must be respected. The media available and the costs of using them vary widely; personal selling to distributors needs sensitive management. Consumer attitudes to price promotions need careful consideration: for example, everyone in the world likes a bargain, but a straight product-related price discount in some cultures indicates inferior quality. Finally, it should go without saying that any promotional material must be in the right language(s) and appropriately designed for the target audience (Plate 21).

CONCLUSION

In increasingly competitive, congested markets, export marketing offers opportunities for sustained business growth by reaching new customers and optimising production factor utilisation. Existing products may target customer needs non-existent in the home market, or command a higher price because they enjoy the status of a new product in the export market. Producer organisations are active in pursuit of this strategy, and individual businesses may pursue it by using an export agent and government-backed export organisations. Exporting nevertheless carries a high risk because it is always vulnerable to interruption by external events beyond management control, so domestic markets and customers must not be neglected in the drive for exports.

CHAPTER 14

Implementation and control

Long-run success in a changing, competitive marketplace has been shown to be closely related to business analysis and planning, but quality implementation is also essential to ensure that what is planned is actually achieved. There are two essential considerations in designing an effective implementation system: the formality of the structures needed, and the implementation tactics. Where control is concerned, the over-riding performance parameters and subsidiary ones need to be distinguished.

IMPLEMENTATION

Large companies necessarily have formal planning and implementation structures which are rarely necessary in the smaller business. However, even the smallest business needs some form of scheduled management system or 'critical path' to ensure that information, inputs and products are made available in time for subsequent activities to be undertaken. This should be self-evident – particularly to farmers who are accustomed to managing production systems which often work to a tight schedule and need effective integration. The complexity of managing even a small-scale and relatively simple marketing system is commonly underestimated, however, and effective management measures are often only set up 'on the hoof'. An undertaking of this kind also frequently overstretches the resources of an owner-managed or family business which simultaneously supplies the production and management team, research division, marketing manager, distributor, advertising manager and PR agent.

Many marketing activities are carried out in parallel: for example, work on advertising and distribution will often be done simultaneously with product testing, plant/equipment design and installation. A clear plan with clear deadlines and effective procedures to integrate the activities and deliver on schedule is essential to cope with this, but this may be all the formality that is required. Sufficient flexibility (*contingency*) must also be built in at every stage to allow for outside factors as well as for creativity and adaptive response within the planning framework.

The tactical implementation of the plan may take either of two forms: piecemeal or 'big bang'. Each is appropriate for different purposes and in different circumstances, and what is the appropriate approach may only be determined in each individual situation.

Piecemeal implementation

Most businesses have a marketing plan in place to which piecemeal modifications are made as and when they are required by new internal objectives and external factors which are likely to include:

- product modifications or new product development
- changes to prices either forced by cost increases, or the need to raise profit margins or appeal to different market segments
- new appeals via advertising and provision of different types of information

- modifications to distribution systems resulting from structural change in the industry, de-listing of products by retailers, need to appeal to new target group, etc.
- competitor activity.

Programme modifications are commonly motivated by the marketing activity of competitors: a price cut, promotion, product enhancement or a new product introduction. The small entrepreneurial business may have a considerable advantage in responding to such environmental changes, because its decision-making procedures and the implementation of any changes should be more rapid than those of a larger competitor firm with a divisional or hierarchic structure. However, no business should abandon a well-planned, long-run strategy simply to respond to the short-run activities of competitors or short-term environmental disturbances.

In order to be effective, any response must also be well-conceived and planned in relation to clear, stated marketing objectives which are consistent with the long-run business objectives. Continuous small changes of direction should be resisted since they are very costly in terms of management time, they tend to disrupt the whole business, and may give every appearance to the market that the business does not know where it is going. In addition to being well-planned, piecemeal changes to the marketing programme therefore need to be presented to the market as part of a sustained programme of continuous product/market development.

The consequences of one small change on the rest of the marketing programme can be much larger than is immediately apparent. The simple introduction of a new flavour to an existing range of farmhouse yogurt, for instance, may require changes to advertising support and discussions with distributors about the appropriateness of the change: for example, its multipack implications (customers like an established flavour combination); listing; stock levels; merchandising support material; possible need for RPOS sampling; price promotion, and so on. All this requires much more management time than is generally appreciated – for the distributor as well as the producer, and this is where the value of good customer research and management of the distributor link becomes apparent, both at the operational level and in evaluating whether the change is worthwhile.

In some business sectors there are conventional times for introducing changes to the marketing mix – the most obvious being the accommodation and leisure sectors, where spring and autumn promotional campaigns are often linked to 'new, improved' product features. Retailing has its own 'seasons': Christmas, Easter, national holidays etc., when changes to a product and/or the marketing mix are expected by both distributors and consumers, and are therefore easier to introduce without adverse comment. New products and variants of established products will, similarly, be timed to appear at the appropriate trade show at which publicity and trial are guaranteed (though the competition is also stronger because everyone else is doing the same).

Big bang

The alternative to a gradualist, piecemeal approach is the 'big bang', when a totally new marketing programme is introduced for a business or a product/product range. This simultaneously involves new product launch and new pricing supported by new packaging, advertising, probably some in-store sampling or promotion, and possibly new distribution arrangements (wider geographical coverage, different retailers etc.).

New product launches of this kind are common in certain business sectors where fashionability is a strong market feature (cars and clothes, for instance), and in sectors characterised by strong habitual buyer behaviour (food). The two factors are often correlated: cars, for instance, tend to be as habitually purchased as food (always buy Ford, every two/three years rather than every week), and in both sectors brand loyalty – or inertial commitment – is strong. Repositioning of existing products is also typically 'big bang', since the objective is to transform the entire customer perception of the product and its marketing, which may entail modifications to everything associated with it.

Such comprehensive activity is obviously more expensive than marginal changes to product packaging or a new promotional campaign. It also demands much more detailed, integrated planning to ensure that everything (at distributor as well as producer level) is ready for the launch, and poised to exploit a successful launch. This means that sufficient stocks of product must be readily available for delivery; stocks of leaflets, brochures and other promotional material must be available; sample sizes must be produced, advertising space booked and advertisements ready to run.

The potential benefit is proportionately greater, since a big bang launch will create an instantaneous impression in the marketplace and attract more attention than a piecemeal approach. It is a higher-risk strategy, however, as even world-wide marketing giants have discovered. For instance, *Coca Cola* spent many months of research effort and logistics planning, and a budget of unknown millions, on a big bang re-launch featuring new packaging, but it was a PR disaster when consumers disliked the packaging.

A high-profile launch therefore has the potential to be a high-profile failure. More typically, a well-planned introduction fails simply because the surrounding noise and congestion in the sector overwhelms even the biggest bang, or because external factors intervene. For example, Box 13.1 reports a case where adverse publicity in a congested food product sector demolished a well-planned, good product introduction which good customer research had indicated would succeed.

Careful research and good planning will therefore not guarantee successful implementation. However, the preparation of a coherent plan which brings together all the aspects of marketing management summarised in Figure 14.1 is essential to preserve a sense of direction and at least anticipate factors which are within the manager's control.

EVALUATION AND CONTROL

Continuous evaluation of performance is an inherent part of good marketing management, providing the means to control current activity and a source of product development. In a large business there is generally a hierarchical control structure, with a marketing manager controlling the implementation of the marketing programme, the production manager controlling production, and a management board overseeing financial and overall performance. Each level will have its own measures of success: sales, percentage of target audience aware of the product, cost per unit of production etc. All these contribute to the measurement of overall business performance, but at board level, measures like sales growth, return on investment, profit, and share value will be more relevant to long-run strategy and medium-term organisation.

In the owner-managed business, marketing, production, and management direction are all in the same hands and the owner is the major shareholder; all

RESEARCH/ANALYSIS

OWN BUSINESS
Physical resources
Financial resources
Management
Current products
Product/service areas

ENVIRONMENT
Political/legal
Economic
Social
Technological

MARKET
Product need
Customer needs
Targetable segments

STRATEGY

Objectives | Resource Utilisation | Target Segment | Competitors

PROGRAMME

Product | Price | Promotion | Place

IMPLEMENTATION

EVALUATION

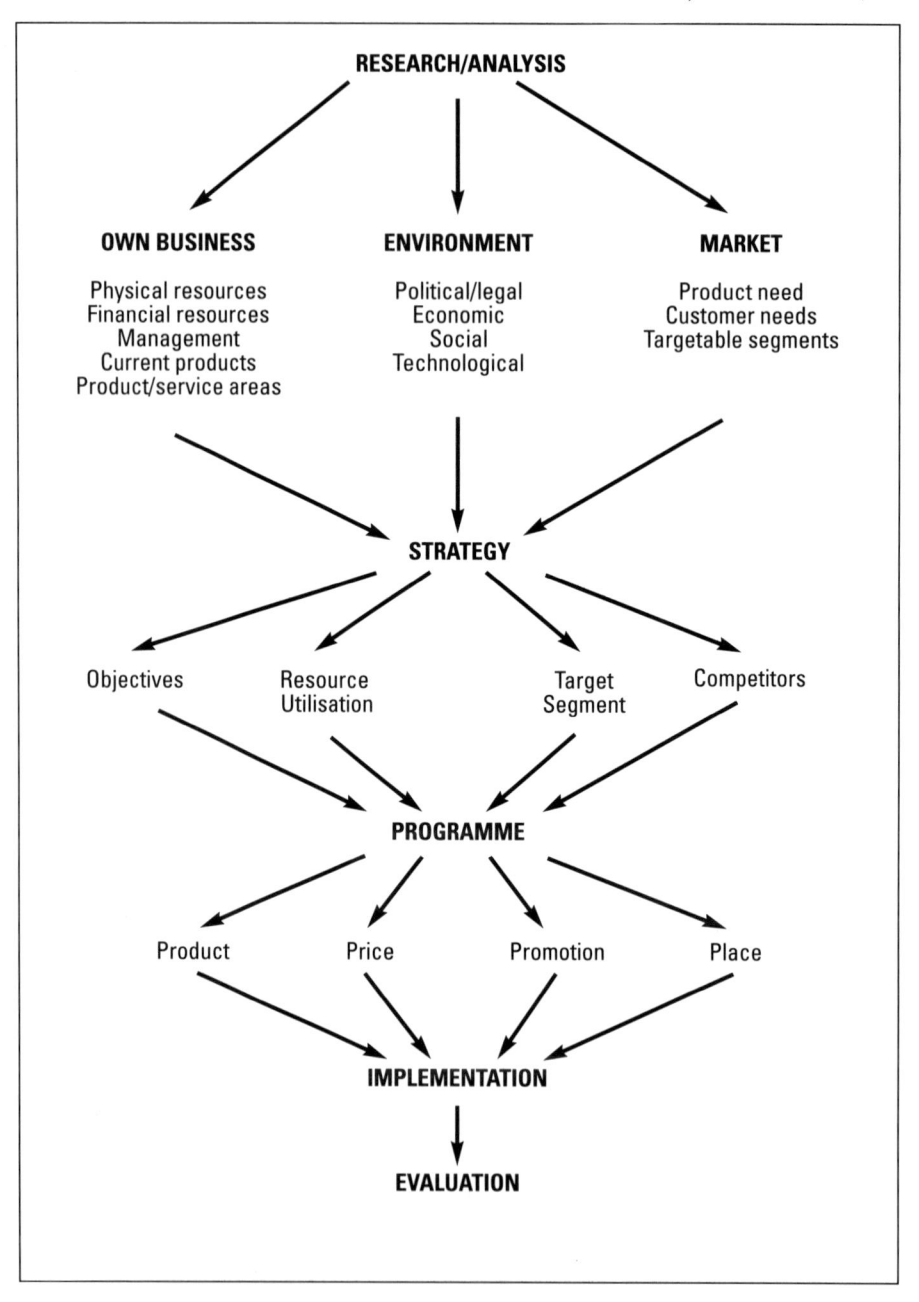

Figure 14.1 The marketing planning process

these considerations are therefore brought together, and are relevant both to operational decisions and strategic planning.

Success must also be measured in terms of objectives set, and performance ideally measured against all the objectives listed in Table 3.1. In practice the data requirement and the analysis which this would require is probably unrealistic, so some key 'headline' control variables should be identified. The primary one must be the reason for undertaking the marketing programme in the first place. For example, if the objective was to create employment, the appropriate measure is how many jobs were created, or how effective was the employment for the personnel involved. If the objective was to raise business profitability by X per cent, the figure achieved is the appropriate measure.

The cause of any variance from the target (positive and negative) should be identified. If the objective was to raise profitability by 20 per cent and the figure achieved is only 10 per cent, is it due to a failure to increase revenue or to control costs? If it is revenue-related, is this the fault of the business (for example, failure of a new product or an advertising campaign) or the result of an environmental change which has affected every other business in the same sector (economic recession)? To explain the reason for under-performance is not to correct it, however, so once it is identified, appropriate and effective remedial action must be taken.

If the problem is environmental, the business may not be able to respond directly to the challenge, but it may seek to mitigate the effect by modifying the marketing plan. In a recession the sensible response may simply be to reduce prices in order to maintain sales volume and (say) a skilled workforce, or make a contribution to fixed costs. A lower profit is accepted in order to stay in business, in anticipation of future recovery and growth in the market. Alternatively, a discount version of a product may be introduced as a short-term response, allowing customers to trade down, while retaining the premium product to which they can return when the recession ends. Conversely, the business may decide to change its objectives and reduce its profit aspirations.

Some businesses normally apply the latter policy, setting objectives at a level they can be *certain* of achieving, and this may be sensible for an individual or a business which is risk-averse. Low-level aspirations may not challenge the business sufficiently, however, and may encourage under-performance. As a general rule it is therefore better to set performance targets slightly higher than are considered attainable, but still within a realistic range, since this strategy encourages that little bit more effort which may make all the difference.

Eventually, the revenue received from the market will form part of a cash flow statement, a trading (profit and loss) account, and a balance sheet, together with the costs incurred in achieving that level of revenue. Only at this stage will it be possible to assess whether the marketing plan has improved the financial viability of the business. Improved financial viability may not come quickly, however. Better marketing performance achieved through higher-quality products or services, new innovative products and better distribution, will take time to show results. Most new supermarkets do not begin to show profits in less than 18 months; some new products only begin to show real profits after 2–3 years, and years may be necessary to pay off the investment in new plant and product launch costs.

CONCLUSION

Marketing is not, in other words, a quick-fix, low-cost, low-management solution to long-run problems of non-viable businesses. It is a *potential* solution, but only when it is combined with investment of resources – financial and managerial. It is therefore not a solution to be undertaken as a last resort, when everything else has failed.

An already strong business which adopts a market-led management approach will in time achieve measurable results, while one which does not will probably slide into a progressive decline because the market will determine revenue and long-run viability. Farm businesses may escape the full impact of this if their incomes continue to be supported by government intervention of one kind or another. For anyone who wishes to achieve a degree of independence and income control, however, the need for better marketing of farm-related output has never been more genuine nor more urgent.

GLOSSARY

Advertising Any paid form of non-personal promotion by an identified sponsor.

Augmented product A standard product offer enhanced by additional customer services and benefits.

Benchmarking Comparative analysis of a business's products and performance with those of competitors or other firms, to identify weaknesses and ways of improving quality and performance.

Brand A name/sign/symbol/design (or any combination) which identifies a seller's goods or services and differentiates them from the competition.

Break-even pricing (*or* target profit pricing) Setting price to break even on the costs of making and marketing a product, or setting price to make a target profit.

Cash cows Low-growth, high-share products or enterprises which generate a high proportion of a business's revenue.

Channel coordination Agreement and effective action between marketing channel members to improve channel efficiency and achieve shared objectives.

Competitive advantage An advantage over competitors achieved by offering customers better value, through lower prices or added benefits which justify higher prices.

Competitor analysis The identification of key competitors, their objectives, strategies, strengths and weaknesses, and likely reaction to a competitive offer.

Convenience store A small store in a residential area or travel location (petrol station, rail/bus station etc.), open long hours seven days a week, with a limited range of high-turnover convenience goods.

Conventional distribution channel A channel consisting of one or more independent producers, intermediary buyers and retailers.

Core product The core benefits which are the minimum consumers expect when buying a product.

Cost-plus pricing Adding a standard mark-up to the cost of the product.

Cultural environment A society's basic values, perceptions, preferences and behaviour patterns, which influence business activity, buyer behaviour and product choice.

Demand curve A curve showing the number of units the market will buy over a given period at different prices.

Demographics Population information in terms of size, density, location, age, sex, ethnic origin, occupation etc.

Derived demand Demand which derives from end use but is indirectly experienced by raw material suppliers, via downstream marketing intermediaries.

Differentiated marketing A marketing strategy which seeks to identify and supply different market segments with distinct, targeted marketing offers.

Direct marketing Direct marketing to the consumer, via a rapidly expanding range of media (mail, Internet, telephone cold calling etc.).

Diversification A business strategy which seeks growth by offering new products to new markets.

Distribution Physical and communications activities involved in the transfer of products and services from point of production to consumption.

Economic environment Economic and financial factors which affect business management and consumer spending.

Elastic demand Total demand for a product which is easily affected by price changes, usually because there is ready availability of substitute products.

Fixed costs Costs which do not vary with production or sales level.

Franchise A contractual association between a producer, wholesaler or service organisation (the franchiser) and independent businesses (franchisees) who buy the right to supply products/services in agreed conditions and (generally) a standard format.

Gross margin The difference between direct product costs and sales revenue.

Horizontal coordination Collaboration between participants at the same level of the marketing channel (inter-producer, inter-processor) to achieve greater efficiency and/or bargaining power, or to add services to the product.

Industrial buyers Individuals and firms purchasing products for processing or as inputs to their business activity.

Inelastic demand Total demand for a product which is relatively unaffected by price changes, especially in the short run.

Institutional market Schools, colleges, hospitals, nursing homes, prisons, armed services etc., which buy goods and services.

Lifestyle A pattern of interests, attitudes and activities which reflect an individual's personality and socio-cultural environment, and influence product purchase.

Logistics (*or* physical distribution) The planning, implementation and control (physical and financial) of activities necessary to deliver materials and products from point of production to consumption.

Macro environment Environmental factors which affect business activity and consumer behaviour but are beyond management control (natural, political, economic, technological and cultural forces).

Market A set of actual and potential customers for a product or service.

Market development A business strategy which seeks growth by identifying and developing new segments and markets for existing output.

Market penetration A strategy which seeks to increase sales of existing products to current market segments by expanding product usage, and by winning customers from competitor offers.

Market-penetration pricing Setting a low price for a new product in order to attract large numbers of buyers and gain a large market share.

Market positioning Identifying and delivering a marketing mix which clearly differentiates a product in the consumer's mind relative to competitor offers.

Market research The collection and analysis of information about customers and markets.

Market segment A group of customers who respond in a similar way to a given set of marketing stimuli.

Market segmentation Dividing a market into distinct groups of buyers with different needs, characteristics or behaviour, who might buy products which target these factors.

Market targeting The process of identifying, evaluating and determining a market at which production and marketing will be targeted.

Marketing audit A systematic appraisal of a business's objectives, strategies and marketing environment, to identify weaknesses and evaluate opportunities, and serve as the basis for revised management strategies and action.

Marketing channel A set of interdependent businesses/organisations which make a product or service available to intermediary and end users.

Marketing environment Factors and forces external to the business which affect the manager's ability to develop and deliver an effective marketing offer.

Marketing management The design, implementation and control of an effective marketing programme, based on preliminary analysis of marketing opportunities and internal capability.

Marketing mix The combination of controllable variables which a manager can modify to deliver a product which satisfies a specified demand: product, price, place and promotion.

Marketing planning process The process of analysing marketing opportunities, selecting target markets, and developing a marketing mix and a management plan to direct the marketing effort.

Marketing research The collection and analysis of information about the marketing environment, with particular reference to competitors, distributors and suppliers.

Marketing strategy A long-run business plan with clearly defined marketing objectives which reflect overall business objectives, internal business capability and environmental conditions.

Media Communications channels which transmit product information to potential consumers: print media (newspapers, magazines, direct mail); broadcast and electronic media (TV, radio, the Internet); display media (public signs, posters, billboards; in-store merchandising).

Merchandising Retail point of sale (RPOS) display and other devices used to attract customer attention; the activities necessary to achieve this (hence supplier 'merchandising' = responsibility for maintenance of RPOS display, product management etc.).

Micro environment The environmental factors which are normally susceptible to management control via the marketing mix, viz., customers targeted, marketing channels used, competitor and supplier management.

Monopoly A market in which there is a single seller: it may be a government monopoly, a regulated private monopoly, or a non-regulated private monopoly.

Natural environment Natural resources and forces which facilitate and/or obstruct production and marketing.

Net profit The difference between the income from goods sold and all expenses incurred.

Net worth The value of a business once all its liabilities have been discharged.

Non-tariff barriers Non-monetary obstacles to cross-border trade: e.g. product standards, health and safety regulations, animal/child/worker welfare standards, political acceptability etc.

Oligopolistic competition A market in which there are few sellers, who are highly sensitive to each other's pricing and marketing strategies.

Opinion leaders (*or* trend setters) People within a reference group who exert influence on others, and may be targeted by a business to encourage product acceptance/higher sales.

Personal selling Person-to-person product presentation, with the object of making sales.

Political environment Political and legal factors which influence business activity and consumer behaviour: government agencies, political parties, pressure groups, laws, legal structures etc.

Portfolio analysis A management appraisal tool which identifies and evaluates the relative contribution to a business of its different products/enterprises.

Price elasticity A measure of the sensitivity of demand to changes in price.

Price skimming Setting a relatively high price for a new product to target the small market segment which can afford it, then progressively reducing the price to appeal to the next segment, and so on down the scale.

Primary data Information collected by/on behalf of the data user for a specific purpose.

Product A physical good, service or an idea, or any combination of the three, which may be offered to satisfy a customer/market want or need.

Product development A business strategy which seeks growth by offering modified or new products to existing market segments.

Product life cycle (PLC) The varying levels of sales and profits over a product's lifetime, which show four main stages: introduction, growth, maturity and decline.

Product mix The combination of product attributes which a seller offers for sale to buyers.

Product position The way in which a product is perceived by customers relative to competitor product offers.

Product specification The combination of technical and other product characteristics necessary to supply a given demand.

Promotion Communications activities necessary to inform customers that products are available for purchase and to persuade them to buy, by stressing their 'unique selling proposition'.

Promotion (*or* communications) mix The combination of paid advertising, non-paid publicity, personal selling and sales promotion which a business uses to communicate with its target markets.

Psychographic segmentation Dividing a market into different groups based on social class, lifestyle or personality characteristics.

Public relations The development of good customer and public relations and a 'corporate image' which triggers customer recognition, and contributes to business viability by generating confidence in the supplier.

Publicity Activities to promote a company or its products by generating public attention through media announcements, news articles etc. (In the past unpaid publicity was sharply distinguished from paid advertising, but the distinction is being eroded as publicity increasingly comes with a price.)

Pull strategy A promotion strategy which uses advertising and sales promotions to build up consumer demand, to which retailers and wholesalers respond by seeking out supplies.

Pure competition A market in which many buyers and sellers are present, and no single buyer or seller has much effect on the going market price.

Push strategy A promotion strategy which uses trade promotions and sales personnel to 'push' supplies to retailers, who promote the product to consumers.

Qualitative research Research which seeks to identify customer motivations, attitudes and behaviour which influence demand.

Quantitative research Research which generates numerical data from a sufficient volume of customers to allow statistical analysis.

Relationship marketing A constructive relationship between marketing channel participants which seeks mutual benefit through regular collaboration, shared objectives and a shared idea of good business practice.

Retailing The sale of goods or services direct to consumers.

GLOSSARY

Sales promotion A short-term incentive (usually but not necessarily price-linked) to encourage sales by intermediary sellers and purchase by consumers.

Sample A segment of the population selected for marketing research to represent the population as a whole.

Seasonality A recurrent, consistent pattern of production/sales movements within the year.

Secondary data Information which already exists somewhere, which can be drawn on for market research in a particular situation.

Service An activity or benefit offered for sale which is essentially intangible.

Speciality store A retail store which carries a narrow product range with great product depth (butcher, greengrocer, cheesemonger).

Straight rebuy A buying situation in which the buyer routinely re-orders a product without any modifications.

Strategic planning Long-run planning to achieve sustained business viability, by identifying clear, attainable objectives and a sound business portfolio, together with responsiveness to environmental change.

Survey The collection of primary research data by interviewing people about their knowledge, attitudes, preferences and buying behaviour.

SWOT analysis A systematic appraisal of current business performance, focusing on internal strengths and weaknesses and external opportunities and threats.

Target market A set of buyers sharing common needs or characteristics which a supplier may decide to serve.

Tariff A tax levied by government against certain imported products, to raise revenue or protect domestic firms.

Technological environment Advances in technology which create new product and market opportunities and affect the operational and financial feasibility of production.

Tele-marketing A direct sales approach to customers made by telephone.

Test marketing The stage of new product development where the product and marketing programme are tested in a setting which approximates real-world conditions.

Total costs The sum of the fixed and variable costs for any given level of production.

Trade promotion Sales promotion directed at wholesalers and retailers to gain their active support for product sales (discounts, allowances, free goods, joint promotion, trade shows etc).

Undifferentiated marketing A marketing strategy which decides to ignore market segment differences and supply a single product via a single marketing mix, usually in the effort to exploit economies of scale and/or efficiency advantages.

Unique selling proposition (USP) The unique benefit which a given marketing offer supplies, which may reflect functional superiority, a high service level, high product quality, lowest price, most advanced technology etc.

Variable costs Costs which vary directly with the level of production.

Vertical coordination Voluntary collaboration between channel participants to achieve greater channel integration and deliver mutual objectives (lower costs, regular trading relationship, higher price etc.).

Vertically integrated channel A channel composed of a single member who performs all the marketing functions from production to retail point of sale.

INDEX

Advertising, 180–181
  above/below the line, 180
  direct and direct response marketing, 180–181
  selection of right medium, 180
  timing, location and presentational format, 180
  trade fairs, 181
  see also Communication process
Agricultural policy
  competitive structure, 41
  and market environment, 35–38
  product prices, 37
  production method, 37–38
  quantity produced, 36–37
  restraint, 36–37
Agrifood marketing system, 52–70
  distribution, 58–61
  retail, 59–61
  see also Distribution of products: Retail (retailer): and headings
  distributors, 53
  firsthand buyers, 53
  industry structure, 55–56
  processing, 54–55
  demand pressures, 54–55
  future trends, 57
  processors, 53
  range, 52
  raw materials, monopolies, 57–58
  see also Raw materials
  small processor sector, 57
  value of products, UK, 1994, 53
AIDA model
  applied to communication message, 178
  applied to effectiveness evaluation, 184, 185
  applied to product presentation, 162
Analysing the current business, 15–31
Analysing the market, 71–94
  see also Consumer; Customer; Market research; Purchasing decision (customer's)
Analysis, product/business portfolio, 21–24
  Boston Box, 21–22
Animal feed, employment and sales, 56
Arrum, new product, 1996, 47
Asia, marketing opportunities, 3
Attitude group research, 79
Auction market, payments, 139, 144
Auction selling, 140–144
  auctioneers' responsibilities, 144
  carcass weight and quality by ultrasonics and image analysis, 142–143
  electronic, advantages, 141–142
  identification of animals, 142
  price determination, 141–142
  quality assessment, 142–143
  size and location of auctions, 141–142
  telephone, 140–141
  terms of trade, 143–144
  payment period, 144
  seller's advantages, 143–144
  traditional, animals to market, 140, 142
  weaknesses, 140–141

Bankruptcy payments, 151
Bankruptcy and unprofitable trading, 117
Bargaining
  distributive, 191
  integrative, 191

Bargaining groups, 190–191
Beef levies to fund promotion, 172
  generic promotion, 171–173
Binding supply agreement, 65–66
Biological factors, effect on food preferences, 78
Boston Box, 21–23
Boston Consulting Group, 21
Bramley apples, 100
  1980s campaign, 99–100
Brand, establishing and maintaining, 28–31
Brand identity and promotion, 172–173
Brand loyalty, 92–93
Brand names and export markets, 212
Brand names and higher pricing, 125
Branding products, 99–100
Branding and promotion, 171–173
Bread, own-label penetration, 61
Bread and wheat-based products, promoting, 107
Bread/biscuits, employment and sales, 56
Break-even chart for determining target price, 124, 124–125
Brewing, employment and sales, 56
Budgeting for promotion, 183
Burger, meals consumed, statistics, 68
Business capability, *see* Business portfolio analysis
Business capability appraisal, 16–32
  current marketing performance, 16
  how good are we at what we do?, 20–21
  marketing strategy, 16
  objectives, 16–19
  resources available, 19–20
  what business are we in, 16
Business environment, 2–3
  finding markets, 2–3
  producers must comply with consumer demands, 2
  see also Consumer
Business, farm
  need for reappraisal, 10
Business forms and the law, 40
Business objectives, clarifying, 11, 12
Business organisation, retail, 60
Business portfolio analysis
  market segmentation, 27–28
  marketing strategies, 25–27
  product width and depth, 22–24
  what business could we be in?, 24–25
Business security and a deficit trading situation, 117
Business, strengths and weaknesses, 15
  see also Business capability appraisal: Farm business
Business survival and pricing objectives, 116–117
Business/product portfolio analysis, 21–24
  Boston Box, 21–23
Business, *see also* Market research: Segmentation, market
Buyer-seller negotiation, 138–152
  auction selling, 140–144
  see also Auction selling
  forward contract, 144–145
  payment periods, 139
  private treaty, disadvantages, 138, 139
  terms of contracts, 147–152
  see also Contracts
  terms of trade, 139–140
  use of agent/professional trader, 139–140

Buyer types, 162–164
  amiables, 164
  analytical, 163
  drivers, 163
  expressives, 163
  see also Customer: Industrial buyers
Bypassing distributors, 58

Campaigns and branding products, 99–100
Carcass classification, 108
Cash cow products, 21, 22
Caterers (catering)
  catering sector as food service provider, 59, 66–70
  demand, 68–70
  institutional, 66–67
  meals outside home, statistics, 67–68, 85
  non-profit ('cost') sector, 66, 67
  profit-making sector, 66–68
Chicken(s)
  export success, 214
  free-range, production, EU law, 40
Children and purchasing decisions, 76
Chilled food, own-label penetration, 61
Chilled ready meals, own-label penetration, 61
Chocolate, own-label penetration, 61
Citation agencies, 184
CMA(Germany), 187, 202
Collaborative promotion, 183
Commodity marketing, 27–28
Commodity marketing board, 187
Common Agricultural Policy
  1992 reform, 1
  quantitative restrictions, 36–37
  see also Agricultural policy
Communication process, 173–185
  cost and budget, 183
  execution and effectiveness, 184
  flow chart, 174
  medium, 179–183
  message, 176–179
  objectives, 175–176
  professional agent, 184
  target audience, 175
Competition-oriented pricing, 119–122
Competition and penetration pricing, 126–127
Competition and price falls, 116
Competitive pricing, 28, 61, 127
Competitors, 46–48
  consumer expenditure, 48
  existing, 48
  horizontal coordination, 47
  inter-firm competition, 46–47
  inter-product competition, 47–48
  managerial control, 46–48
  near-identical products, 47
  potential, 48
Computer home shopping, 64–65
  multiples' reaction, 64–65
Computer and TV home shopping as direct selling, 166–167
Confectionery, employment and sales, 56
Consultants, marketing, obtaining advice, 14
Consumer awareness of food quality, 73–74
Consumer data, and market research, 86
Consumer demands and complying of products, 2

228

INDEX

Consumer food markets, characteristics, 27–28
Consumer groups and ethical marketing, 44
Consumer markets
 segmentation, 28–31
 see also Segmentation, market segmentation variables, 90–93
Consumer opinion, 44
Contracts
 forward, 144–145
 market-specifying, 145–146
 production management, 146
 resource-providing, 146–147
 terms, 147–152
 fairness and sustainability, 151
 price, 148–150
 basic formula pricing, 148
 pricing by tender, 149–150
 system pricing, 148–149
 quality assessment and remuneration, 150
 quantity, 150–151
 security of payment, 144, 151
 specific negotiation, 147
 terms of trade, 151
Control of products, 219–221
Cooperative marketing, 187–188
 see also Group marketing
Cooperative organisation, 194–199
 cooperative business objectives, 196–197
 finance, 198–199
 kinds of organisation, 196–197,
 management structure, 198
 membership commitment, 194–195
 organisational structure, 195
Cooperative principles, 191–194
 democratic control, 192–193
 financing according to use, 193–194
 open membership, 192
 return to capital, 193
 service at cost, 193
 voluntary membership, 192
Cornish growers promoting a regional identity, 189
Cosmetics and toiletries, own-label penetration, 61
Cost-based pricing, 122–123
Costs and differentiated marketing, 28–31
 see also Differentiated marketing
Costs of promotion, 183
Cox apples, 1980s campaign, 99
Cultural factors, effect on food preferences, 78
Currency exchange rates for export marketers, 42–43
 fixed exchange rates, 42–43
 floating exchange rates, 43
Customer, changing perceptions, 106
Customer preferences unknown, 127
Customer satisfaction and needs, 8–9
Customer service and
 company/store image, 64
 see also 'Shopping' environment
Customer service questionnaire, 23
Customer, understanding, 71–79
 added value, 71, 72
 after-sales, 71
 functional product characteristics, 71–72
 local dealer, 71
 price discount, 71
 purchasing decision, 72–76
 see also Price: Purchasing decision (customer's)

Demand (customer)-oriented pricing, 124–125

Demand-led not production-driven marketing, 9
Demographic features and the export market, 208–209
Demographic gaps, 105
Demographic segmentation, 91–92
Differentiated marketing, 28
 business strengths and weaknesses, 29–30
 degree, 28–31
 higher marketing costs, 29–30
 relative costs, 29
Direct marketing, 167–168
 of wine, flowers, gift food packs, 65
Direct response coupons delivered to the door as direct selling, 166–167
Direct selling, 166–167
Directional objectives, 18, 19
Distribution decisions, 135–152
 and business objectives, 136
 buyer-seller negotiation, 138–152
 see also Buyer-seller negotiation
 criteria for channel selection, 137–138
 firsthand intermediaries, 137–138
 interaction with market mix, 135–136
 marketing raw materials, 137–138
 milk and raw materials ex-farm, 136
Distribution and the export market, 214
Distribution management, 164–166
 batch deliveries, 165
 customer service, 165
 electronic data interchange, 164–165
 inventory control, 164
 logistics management, 164–166
 transportation, 165–166
Distribution of products, 58–61
 bypassing, 58
 catering sector, 59
 functions of distributors, 58–59
 range of activities, 58
 retail, 59–61
 retailer marketing strategies, 61–64
 wholesalers/retailers, 58–59
Distributor selection, 155–164
 channels, 154–155
 follow-up selling, 164
 gaining access, 160–161
 identifying alternatives, 156
 management, 156
 product presentation, 161–164
 refining the selection, 156–160
 selecting types and gaining access, 156, 160
Diversification
 opportunities, 5
 portfolio analysis, 25, 26
Dog products, 21, 22,
Doorstep milk delivery, product design, 96–97
Drink, total expenditure, UK, 1994, 53
Drink products, value, UK, 53
Drinks, soft, employment and sales, 56

Eastern Europe, marketing opportunities, 3
Eating out
 future trends, 69
 see also Meals outside home
Economic development and the export market, 207–208
Economic environment, see Environment, economic
Education levels and markets, 44
Educational factors, effect on food preferences, 78
Elasticity of pricing, 120–122
 relationship of total revenue to, 120–121

Electronic data interchange and distribution, 164–165
Employment by UK food industry, 55–56
Environment, economic, 41–43
 competitive structure, 41
 international politics, 41
 price makers, 41
Environment, macro, 35–45
 agricultural policy, 35–38
 see also Agricultural policy
 currency exchange rates, 42–43
 economic environment, 41–43
 environment and rural policy, 38–39
 legal environment, 39–40
 socio-cultural environment, 43–44
 technological environment, 45
 trading access, 42
 wider political environment, 39
Environment, marketing, 33–50
 exploiting, 33
 legal, 39–40
 macro, definition, 33–35
 major factors, 33
 micro (task), definition, 33–35
 opportunities and threats, 15
 rural policy, 38–39
 wider political, 39
 see also Agricultural policy; Socio-cultural environment; Technological environment
Environment, micro, 46–50
 competitors, 46–48
 suppliers, 48–50
Environment, 'shopping', 60
Environmental constraints, identify, 11, 12
EPOS (Electronic Point Of Sale)
 data, 64
 and retailing costs, 62
 scanning, 86
Equilibrium price, 119–120
Ethical marketing, 44
Ethnic meals, meals consumed, statistics, 68
Eurosceptic views and the export market, 206
Evaluation
 and control of products, 219–221
 financial viability proven, 221
 identifying variance from target, 221
 low-level aspirations, 221
 maintaining targets, 221
 measure of success/failure, 221
'Exotic' foods, increase, 69
Export marketing, 201–215
 additional costs, 201–202
 EU, common market, 203–204
 export process, 203–209
 government intervention, 202
 market entry, 210–215
 marketing programme, 211–215
 planning, 201–204
Export process
 country profile, 207–209
 customer selection, 209
 trading policy, 204
Export restrictions, 206–207
External objectives, 17, 18f

Farm business
 limitations, 95
 market research, 79–87
 see also Market research: Segmentation, market
 maximising revenue, 104–109
 new product development, 110–112

229

product decision, 95–113
product life cycle, 101–104
  see also Business: Product decision
Farm shop, product width and depth, 22–24
Farmers
  early relations with consumers, 1
  change with competition, 1
  change with removal of government support, 1
  cutback of subsidies, 1
  need for increased efficiency, 3
  see also Producers
Farmhouse accommodation, 96–97, 98–99
  decisions, 96–97
  elements of a product, 96
  uses, 81
Farmhouse catering (ice-cream parlours and milk bars), 70
Farmhouse processors
  finding new ideas, 81–82
  see also Market research
  new products, 81
Fast food catering and franchising, 168
Fast-food restaurants, 69–70
Fish and chips, meals consumed, statistics, 68
Fish processing, employment and sales, 56
Fixed exchange rates, 42–43
Floating exchange rates, 43
Follow-up selling, 164
Food consumption data and market research, 83–85
Food faddist, seven types, 93,
Food From Britain, 187, 202, 214
Food industry, UK
  future, 57–58
  small processors, 57
  structure, 55–56
  see also Agrifood marketing system
Food preferences, factors influencing, 78, 79
Food processing, see Processing
Food products, value, UK, 53
Food quality, consumer awareness, 73–74
Food (total), own-label penetration, 61
Food, total expenditure, UK, 1994, 53
Formula pricing, basic, 148
Forward contract, 144–145
Franchise selling, 166, 168
Free range chicken production, EU law, 40
Free trade, economic environment and the EU, 42
Freephone invitations as direct selling, 166–167
Frozen food, own-label penetration, 61
Fruit processing, employment and sales, 56
Futures markets, 130–134
  description, 130–131
  how they work, 131–134
  speculation, 134
  target-level profit pricing, 133–134

GATT, 1994, effect on producers, 2
GATT agreement requirements and the EU, 42
Generic advertising, 171–172
  British beef, 172
Geographic segmentation, 91
Geographical gaps, 105
Government controls, see Agricultural policy; Environment, macro; Environment, micro; Legal regulations

Group marketing, 187–199
  bargaining groups, 190–191
  cooperative organisation, 194–199
  cooperative principles, 191–194
  objectives, 188–189
  operating groups, 189–190
Groups influencing food preferences, 77–79
Growers' identity is revealed, 173

Health and safety and export marketing, 205–206
Hedging in futures markets, 132–133
HGCA (Home Grown Cereals Authority), 187
  enterprise awards, 81
Home shopping by computer, 64–65
  multiples' reaction, 64–65
Hotels
  meals consumed, statistics, 68
  and small suppliers, 70
Household goods, own-label penetration, 61

Ice-cream, employment and sales, 56
Identify current strengths and weaknesses, 11, 12
Image analysis and the auction market, 142–143
Implementation, 217–219
  big bang, 218–219
  introducing changes, 218
  new products, 218–219
  piecemeal, 217–218
  programme modifications, 218
Import control, quantitative, 205
Import taxes, 204–205
Income, maintaining, and diversification, 5
Industrial buyer, types, 162–164
Industrial buyers
  buying process, 74–75
  and consumers, 74
  long-run trading, 74
  see also Buyer Information:
  marketing, 13–14
  analysis, 13–14
  interpretation, 13–14
  research, 13–14
Input price rise effects, 49–50
  factors involved, 49–50
Inter-firm competition, 46–47
Inter-product competition, 47–48
Internal objectives, 18
International Food and Drink Exhibition, 181
International politics and agricultural policy, 39, 41
Internet ordering, 65
Introductory price offers, 126–127
Inventories of stock and distribution, 164, 165

Labelling regulations and the export market, 206
Lamb mince, rise in sales, 80
Legal regulations
  concerning food production, 38
  EU laws, 40
  and marketing environment, 39–40
Legal requirements and the export market, 215
Literacy levels and marketing, 44
Logistics and distribution management, 164–166
Long-run trading relationships, 74
Loss leaders, 128

Macro-economic environment, 41
Macro environment, see Agricultural policy: Environment
MacSharry reform of CAP, 1
Mail order as direct selling, 166–167
Management expertise, value, 19
Management, how good are we at what we do?, 20–21
  data required, 20–21
Management, marketing, 9–10
  deciding objectives, 9–10
  definition, 8–9
  production v. market-oriented management, 9–10
Management, resources available, 19
Market development, portfolio analysis, 25, 26
Market penetration, portfolio analysis, 25, 26
Market prospects, 24–24
Market research, 79–87
  consumer data, 86
  identifying additional markets, 81–82
  identifying markets, 80–83
  industrial raw materials, 79–80
  interpreting the data, 85–87
  primary data collection, 86–87
  product-by-use matrix, 81, 81–82
  researching the market, 82–87
  see also Segmentation, market
Market segmenting, see Segmentation: market
Market-specifying contracts, 145–146
Marketing
  definition, 8–9
  differentiation from selling, 7
  explaining, 7–14
  importance, 1–6
  see also Customer; headings throughout, especially Business:
  Information: Planning
Marketing challenge, 5
  maintaining income, 5
Marketing consultants, obtaining advice, 14
Marketing environment, 33–50
  see also Environment, marketing
Marketing opportunity, 3–5
  better marketing through existing channels, 4–5
  consumer prices, 3–4
  identify, 11, 12
  on-farm processing and direct sale, 4
  processing and intermediary services, 3
Marketing planning process, 219, 220
Marketing programme
  devise, 11, 12
  implement and control, 11, 12
  measure and review, 11, 12
  modify and develop, 11, 12
Marketing strategies, portfolio analysis, 25–27
Marketing system
  agrifood, see Agrifood marketing system
  channels, 51–54
  understanding, 51–70
Mass marketing, 27–28
Meals outside home, 85
  future trends, 69–70
  statistics, 67–68
Meat, product/use matrix, 82
Meat & Livestock Commission, 187
Meat advertising campaigns, 98
Meat classification grades, 108
Meat industry, buyers, 56
Meat and meat products, market

INDEX

research, 80–81
Meat products, own-label penetration, 61
Membership agreements (member-producer group contracts), 145–146
Mercasur (South American countries), and the EU countries, 42
Micro-economic environment, 41
Micro environment, *see* Environment
'Middleman' function, 53
Milk marketing 'cooperatives', 195–196
Milk Marque calls for world class UK plants, 188
Milk quality standards, achieving, 109
Milk quota and supply, 56
Milk/milk products processing, employment and sales, 56
Milling grain, employment and sales, 56
Modified re-buy, 164
Multiples
 binding supply agreements, 65–66
 change in size and style, 60
 distributor selection, 158
 and home shopping by computer, 65
 increase, 60
 re-opening small outlets, 65

NAFTA (North American Free Trade Area), and the EU countries, 42
New Covent Garden seeks fresh life, 140
New product development, 110–112
 process, 111–112
 sources for ideas, 110
New product launch, 218–219
New product pricing, 126–127
New products, price level, 118
Niche marketing, 13, 27–28

Objectives, *see* Strategies and objectives
Objectives, business, appraisal, 16–19, 18
 directional, 18, 19
 external, 17, 19
 performance, 17, 19
On-farm processing and direct sale, 4
Operating groups, 189–190
Opportunities, environmental, 15
Organic farming, conversion to, 106
Organic foods
 popularity and production, 38
 superiority?, 98
Origin of product
 buyer's right to know, 143
 revealed, 173
Out of Home Eating Monitor, 85
Outlets, small, by large multiples, 65
'Over-riders', 129
Own-label products, 60–61, 62–63

Packaging design improved, 106
Packaging regulations, 40
Partnerships and the law, 40
Payment
 auctioneers' responsibilities, 144
 period, 160
 see also Contract
Penetration pricing, 126–127
Perceived value pricing, 125
Performance objectives, 17, 18
Periodic disposal sales to reduce stocks, 127–128
Personal factors, effect on food preferences, 78
Personal selling, *see* Direct selling
'Pester power' of children, 7
Pet food, own-label penetration, 61

Petrol station outlets, 65
Physical contracts and the futures markets, 130–134
Physical and service attributes of product, 97
Physiological factors, effect on food preferences, 78
Pizza, meals consumed, statistics, 68
Place, product, price and promotion (4Ps), 12, 13
Place decision (placing the product), 135–152
 see also Distribution decisions
Planning, market/farm directed, 13
Planning process, 10–13
 business plan, 11
 consecutive stages, 11–12
 effective management, 11
 terminology, 10–11
Political factors and agricultural policy, 39
Potato varieties changed for crisps or chips, 106
Presentation of product, *see* Product presentation
Price
 changes, 127
 competitiveness, 61
 falls, and competition, 116
 high, and decrease of sales, 115
 lower, and increased sales, 115
 and the marketing system, 54
 product, promotion and place (4Ps), 12, 13
Price and agricultural policies
 government determination, 37
 support prices, 37
 temporal variations, 37
Price decision, 115–134
 futures markets, 130–134
 see also Futures markets
Price determination at auction, 141–142
Price and the export market, 213–214
 currency exchange rates, 213–214
 introductory phase, 213
Price level, 118–125
 competition-oriented pricing, 119–122
 competitive pricing, 127
 cost-based pricing, 122–123
 demand (customer) oriented pricing, 124–125
 determination, 115
 interacting factors, 118–119
 new product pricing, 126–127
 new products, 118
 price variation, 128–130
 product range prices, 127
 target-level profit pricing, 123–124
 variation around, 115, 125
Price maker, 115
Price rise effects, 49–50
Price skimming, 126
Price structures, economic environment factors, 41
Price taker, 115
Price and terms of contracts, 148–150
 basic formula, 148
 by tender, 149–150
 system, 148–149
Price variation, 128–130
 quantity-related prices, 129
 seasonal/temporal pricing, 129–130
 spatial pricing, 129
Pricing, competitive, 28
Pricing objectives, 116–118
 business survival, 116–117
 current profit maximisation, 117

product quality leadership, 118
 sales maximisation, 117
Processing, future trends, 57
Processing raw materials, 54–55
 demand pressures, 54–55
Producer groups and promotion, 172
Producer Responsibility Obligations (Packaging Waste) Regulation, 40
Producers
 attitudes to 'business', 2–3
 'complying' and CAP, 2
Producers (farmers and suppliers), binding supply agreements, 65–66
Product decision, 95–113
 maximising revenue from existing products, 104–109
 new product development, 110–112
 product design, 95–100
  see also Product design
 product life cycle, 101–104
Product design, 95–100
 branding, 99–100
 physical and service attributes, 97
 psychological attributes, 97–99
Product development, portfolio analysis, 25, 26
Product labels and EU law, 40
Product myopia, 16
Product, own-label, 60–61, 62–63
Product presentation, 161–164
 buyer behaviour, 162
 buyer types, 162–164
 sales interview, 162
 stages in selling process, 161–162
 technical information, 161
Product, price, promotion and place (4Ps), 12, 13
Product range
 and distributor selection, 159
 prices, 127
 retail, 60
 retailer strategies, 63–64
Product source
 buyer's right to know, 143, 173
 revealed, 173
Product standardisation
 difficulties with farm products, 107
 purchase by description, 107
 'quality products sell themselves', 109
 quality standards, 108–109
Product/business portfolio analysis, 21–24
 Boston Box, 21–23
Production costs, 28–31
Production management contracts, 146
Production methods and agricultural policies, 37–38
 enforceable controls, 37–38
 government action, 37
Production v. market-oriented management, 9–10
Promotion, 169–186
 collaborative, 183
 communication methods, 169–170
 communication process, 173–185
 costs, 170
 definition, 169
 and the export market, 214–215
 objectives, 171–173
 product, price and place (4Ps), 12, 13
 professional input, 170
 promotion mix, 170–173
 service products (advertising, TV, etc.), 170
 see also Communication process
Psychographic segmentation, 92
Psychological attributes of physical product, 97–99

attitude and manner of sales personnel, 98
branding, 99–100
lifestyle and image, 98
luxury preparation, 97
organic food, superiority?, 98
packaging, 97–98
professional buyers, 98
Psychological factors, effect on food preferences, 78
Publicity, 181–182
  advertorial copy, 182
  definition, 181
  non-paid, 182
  public relations, 181–182
Pubs/bars/cafes, meals consumed, statistics, 68
Purchasing decision (customer's), 72–76
  alternative products, 74
  analysing buyer behaviour, 76–79
  attitude groups, 79
  consumer psychological and social factors, 77–78
  consumers and industrial buyers, 74
  factors influencing food preferences, 78, 79
  habitual buying behaviour, 72–73
  high and low involvement types of purchases, 72
  pester power of children, 76
  process, 73–76
  relationship marketing (long-run trading), 74
  repeat purchases, 74
  vegetarianism, 76
  who is the customer?, 74–76

Quality assessment and remuneration and terms of contracts, 150
Quality and price ranges, 127
customer preferences, 127
Quantity-related prices, 129
Quantity and terms of contracts, 150–151
Question mark products, 21, 22
Questionnaire, customer service, 22
Quick service restaurant sector, profit catering, 67

Radio advertising as direct selling, 167
Range of goods, increase by retailers, 60
Range of products, see Product range
Raw materials
  competition and monopolies, 57–58
  marketing, 137–138
  processing, 54–55
  processor prices, 116
  value, UK, 53
Reappraisal and reorganisation, 14
Regional factors, effect on food preferences, 78
Regulations concerning food production, 38
Religious factors, effect on food preferences, 78
Research
  marketing, key to management function, 13–14
  techniques, analysis of existing data, 21
Resource allocation decision, 20
Resource-providing contracts, 146–147
Resources, additional, determination, 20
Resources available, utilisation, 19–20

Restaurants
  meals consumed, statistics, 68, 85
  and small suppliers, 70
Retail (retailer)
  decline in food retailers, 60
  distribution, 59–61
  future trends, 64–66
  increase of range of foods, 60
  marketing strategies, 61–64
  own-label products, 60–61, 62–63
  supply links and market alliances, 63
Rural policy and environment, 38–39

Safety and export marketing, 205–206
Sales maximisation and pricing objectives, 117
Sales, periodic, to reduce stocks, 127–128
Sales promotions, 182–183
Seasonal/temporal pricing, 129–130
Segmentation, market, 27–28, 87–93
  consumer interests, 30–31
  continuous monitoring, 88
  decision, 28–31
  differentiated marketing, 28
  segment accessibility, 89
  segment appropriateness, 89–90
  segment identification, 87–88
  segment measurability, 88–89
  household statistics, 89
  segmentation variables, 90–93
  undifferentiated marketing, 27–28
  unserved segment, 30–31
Selling method and distributor selection, 159–160
Selling the product, 153–168
  direct, 166–167
  direct/indirect factors, 153
  distributor selection, 155–164
  distributors, 153–155
  follow-up, 164
  sales skills, 153
  see also Direct marketing; Direct selling; Distribution management; Distributor selection
'Shopping' environment, 60
  future trends, 64–66
  and product range, 63–64
Slaughtering, employment and sales, 56
Small-ad columns as direct selling, 167
SMART, acronym explained, 17
Socio-cultural environment and agricultural policy, 43–44
  demographic factors, 43
  social factors, consumer opinion, educational levels, 44
Socio-economic classification for food consumption, 92
Socio-economic factors, effect on food preferences, 78
Socio-political structures and export process, 208
SOPEXA(France), 187, 202
Source of product
  buyer's right to know, 143
  revealed, 173
Spatial pricing, 129
Special prices, 128
Speculation in the futures markets, 134
Spirit distilling, employment and sales, 56
Star products, 21, 22
Starch products, employment and sales, 56
Storage and seasonal pricing, 130
Stores
  increase in size and style, 60
  'shopping' environment, 60

Straight rebuy, 164
Strategies and objectives, 12–13
  clarify, 11, 12
Strawberry season, shifting, 105
Strengths, business, 15
Supermarket cafeterias, 67
Supermarkets, see Multiples
Suppliers, 48–50
  abandonment of enterprise, 48
  access to supplies, 48
  identifying sources, 48
  input price rise effects, 49–50
  factors involved, 49–50
  order sizes, 48
  see also Producers
SWOT (strengths, weaknesses, opportunities and threats) analysis, 15
System pricing, 148–149

Take-over prospects, weak businesses, 24–25
Target-level profit pricing, 123–124
  in the futures markets, 133–134
Tariffication, 205
Technical advice and legal requirements, 40
Technological environment, 45
Telemarketing (telephone 'cold calling' sales) as direct selling, 166–167
Telephone and electronic business/service directories as direct selling, 166–167
Temporal/seasonal pricing, 129–130
Tender, pricing by, 149–150
Terminology, planning, 10–11
Terms of trade and distributor selection, 160
Threats, environmental, 15
Trade associations and legal support, 40
Trade fairs, 184
  for advertising products, 181
Trading, see Buyer-seller negotiation
Transportation and distribution management, 165–166
TV advertising as direct selling, 167

Ultrasound scanning and the auction market, 142–143
Undifferentiated marketing, 27–28
Unique selling proposition, 178
Unit pricing, 127
US agricultural policy, consequences for EU, 42, 42

Vegetable processing, employment and sales, 56
Vegetable producer in receivership, 139
Vegetarianism increase, 76
Voluntary producer cooperative, 187
Voluntary restraint agreements, 205

Weaknesses, business, 15
Wines, employment and sales, 56
Work place, meals consumed, statistics, 68